Mathematical Modeling and Computational Intelligence in Engineering Applications

Antônio José da Silva Neto
Orestes Llanes Santiago
Geraldo Nunes Silva
Editors

Mathematical Modeling and Computational Intelligence in Engineering Applications

cujae

Editors
Antônio José da Silva Neto
Universidade do Estado do Rio de Janeiro
Instituto Politécnico, Nova Friburgo
Brazil

Orestes Llanes Santiago
Instit. Sup. Polit. José Antonio Echeverría
(CUJAE)
Dpto. de Automática y Computación
Havana, Cuba

Geraldo Nunes Silva
Universidade Estadual Paulista - UNESP
IBILCE, São José do Rio Preto, Brazil

ISBN 978-3-319-81766-8   ISBN 978-3-319-38869-4   (eBook)
DOI 10.1007/978-3-319-38869-4

Printed on acid-free paper

This Springer imprint is published by Springer Nature
The registered company is Springer International Publishing AG Switzerland

To Gilsineida, Lucas and Luísa
Antônio José da Silva Neto

To Xiomara, Maria Alejandra and Maria Gabriela
Orestes Llanes Santiago

To Heloísa, Gabriela, Isabella and Tiago
Geraldo Nunes Silva

Foreword

"Model building is the art of selecting those aspects of a process that are relevant to the question being asked." J.H. Holland[1]

Mathematical modeling has been applied to Engineering for centuries. Undoubtedly it was behind many of the wonders of the industrial age. More recently, we have witnessed the blossoming of many areas where mathematical modeling has an ever increasing impact. Among such areas, we can mention biomedical research, semiconductor technologies, industrial management, chemical engineering, and power systems, to cite just a few.

This volume provides a cross section of a wide range of applications where mathematical modeling techniques can be exemplified in a fairly self-contained form so as to allow a non-expert to get to grips with the problems and the modeling approach.

The chapter on the morphology of abdominal aortic aneurysms (AAA) as predictor of their rupture combines sophisticated computational fluid dynamics modeling of flows in three dimensions with statistical techniques so as to assess the AAA rupture risk.

The chapter on cytoskeleton dynamics deals with an alternative approach for representing the displacement of molecules and cytoplasmic fluid in the extremely narrow and long filopodia (microspikes). The authors discuss strategies to couple the particle-in-cell method with algorithms for laminar flow so as to model the two phases of actin dynamics. Namely, the polymerization into filaments which are pulled back into the cell and compensatory G-actin drift toward its tip to supply polymerization. The models are validated by using experimental data from fruit-fly nerve cells.

The chapter on artificial neural networks in fault diagnosis deals with the difficult task of designing a fault diagnosis system using a limited amount of information.

---

[1]Holland, JH (1995) Hidden Order. Addison-Wesley, New York, USA.

Such data may even be incomplete due to loss of information caused by a number of problems, being the malfunction of the measurement channel one of them.

The chapter on time-dependent incipient fault diagnostics provides an enthralling application of inverse problem theory. Not only it addresses a problem of practical importance, but it also crosses the bridge between the terminology used in Statistics and that used in inverse problems.

The chapter on fault detection with kernel principal component analysis tackles key issues in the detection of abnormal events so as to determine their causes early enough. The theory was applied to a benchmark test case, and the experiments show that the combination of the proposed approach with the kernel method described in the chapter helps to attain high performance rates in the execution of fault detection tasks.

The chapter on chromatography process identification is another example of an important application of modeling to mass transfer mechanisms that makes use of statistical techniques and Markov chain Monte Carlo methods. It also deals with the issue of uncertainty quantification in such approach—in other words, a combination of current mathematical methods with relevant problems.

The chapter on the analysis of a new anomalous diffusion model looks at the inverse problem associated with an interesting formulation of diffusion phenomena with retention proposed by Bevilacqua, Galeão, and coauthors. The model may have impact in the description of population dynamics, chemical reactions inducing adsorption processes, and multiphase flow through porous media. The authors make use of the classical maximum likelihood method as well as two techniques within the Bayesian inference framework, namely, the maximum a posteriori estimator methods and the Markov chain Monte Carlo methods.

The chapter on computational time reduction is dedicated to the determination of physicochemical parameters in a model for adsorption in a chromatographic column. The speedup of the estimation under investigation is a result of the reduction of the order accuracy in the direct problem solution under investigation.

The chapter on the structural behavior of communication towers deals with a classical yet current problem in Mechanical and Civil Engineering. It makes use of integrated wind tunnel experiments and FEM numerical methods in order to study the behavior of certain lattice towers and the effect of the wind load on such towers.

The chapter on preconditioners deals with an application that could not be missed on a modeling volume, namely, oil reservoir simulations. Domain decomposition techniques are combined with incomplete factorization at subdomain level with the ultimate goal of designing scalable parallel preconditioners, which can then be used in simulations.

The chapter on electrical power system reliability displays yet another important application of Monte Carlo simulation (in this case of nonsequential simulation) so as to assess the reliability of generation systems.

The chapter on the simulation of electronic devices presents a new proposal for modeling and extracting the model parameters for polymeric thin film transistors (PTFTs).

The unifying thread in the various parts of the present volume is the impact of the different applications as well as the novelty of the approach to each of the problems under consideration. It is thus a welcome companion for a course on mathematical modeling. Furthermore, each chapter could be considered as a *case* where the authors use the state-of-the-art tools available to tackle current challenges in modern Engineering.

Jorge P. Zubelli
IMPA—Institute of Pure and Applied Mathematics
Rio de Janeiro
Brazil

Preface

In November 2014, as part of the activities for the celebration of the 50th Anniversary of the Ciudad Universitaria Jose Antonio Echeverria (CUJAE), where lies the largest technological University of Cuba, a work team, integrated by professors from CUJAE and IPRJ-UERJ of Brazil, organized the International Symposium of Mathematical Modeling applied to Engineering as part of the XVII International Convention of Engineering and Architecture, in Havana.

After a double peer review process, 65 articles were accepted for oral presentation in the Symposium. Taking into account the quality of the papers presented, the scientific committee of the Symposium decided to propose the edition of a book and performed a second round of selection in order to define those that would be chosen to compose it. The criteria used in this selection were three fundamentally:

1. The quality of the paper, taking into account the evaluation of the reviewers in the first round of review
2. Evaluation of the technical content by the book editors
3. Representation of different areas of engineering applications

Finally, 12 articles were selected, with the possibility of being expanded and published as chapters of the book. They cover applications in automation, biomedical, chemical, civil, electrical, electronic, geophysical, and mechanical engineering, showing the multidisciplinary application of the mathematical and computational modeling tools.

The primary differentiation and distinctive quality of this book is the wide variety of applications, mathematical and computational tools, and original results presented, all with rigorous mathematical procedures.

Furthermore, the authors from five countries and 16 different research centers contribute with their expertise in both the fundamentals and real problem applications, based upon their strong background on modeling and computational intelligence.

The following is a summary of the book's core topics:

- Rupture risk prediction of abdominal aorta aneurysm
- Computational modeling of cytoskeletal dynamics
- Fault diagnosis in industrial systems
- Parameters and uncertainties estimation in chemical processes
- Development of statistical and numerical strategies for computational cost reduction
- Reliability analysis of electrical generation-transmission systems
- Preconditioners for solving simulation problems
- Modeling of polymeric thin film transistors

The editors consider that this book is intended for use in graduate courses of engineering, applied mathematics, and applied computation, where tools as mathematical and computational modeling, numerical methods, and computational intelligence are intensively applied to the solution of real problems.

Nova Friburgo, Brazil   Antônio José da Silva Neto
Havana, Cuba   Orestes Llanes Santiago
São Paulo, Brazil   Geraldo Nunes Silva
July 2016

Acknowledgments

The editors acknowledge the decisive support provided by the Sociedade Brasileira de Matemática Aplicada e Computacional (SBMAC) and the Springer representation in Brazil for the publication of this book. Our gratitude is also due to the Brazilian Research supporting agencies: Fundação Coordenação de Aperfeiçoamento de Pessoal de Nível Superior (CAPES), Conselho Nacional de Desenvolvimento Científico e Tecnológico (CNPq), and Fundação Carlos Chagas Filho de Amparo à Pesquisa do Estado do Rio de Janeiro (FAPERJ), as well as to the Cuban Ministry of Higher Education MES (Ministerio de Educación Superior).

Orestes Llanes Santiago acknowledges specially Conselho Nacional de Desenvolvimento Científico e Tecnológico (CNPq) for the Special Visiting Professor grant No.401023/2014-1.

Acknowledgments are also due to the authors and reviewers, whose expertise were fundamental in the preparation of the book.

Contents

**1   Preliminary Correlations for Characterizing
the Morphology of Abdominal Aortic Aneurysms
as Predictor of Rupture**   1
Guillermo Vilalta Alonso, Eduardo Soudah, José A. Vilalta
Alonso, Laurentiu Lipsa, Félix Nieto, María Ángeles Pérez,
and Carlos Vaquero
1.1   Introduction   2
1.2   Material and Methods   3
    1.2.1   AAA Geometry: Segmentation and Reconstruction   3
    1.2.2   AAA Geometry: Size and Shape Indices   3
    1.2.3   Computational Methods   5
    1.2.4   Statistical Analysis   6
1.3   Results and Discussions   6
    1.3.1   Relationship Between Hemodynamic
        Parameters and Geometric Indices of the AAA   8
1.4   Conclusions   11
References   14

**2   Mathematical-Computational Simulation
of Cytoskeletal Dynamics**   15
Carlos A. de Moura, Mauricio V. Kritz, Thiago F. Leal,
and Andreas Prokop
2.1   Introduction   16
2.2   The Object of Our Study   17
    2.2.1   Actin Dynamics   17
    2.2.2   Filopodia   19
2.3   Suitable Biological Systems for Parallel
    Experimental Approaches   20

2.4   Modeling Filopodia   21
        2.4.1   Previous Models   21
        2.4.2   Can Diffusion Alone Explain G-Actin
                Delivery to the Filopodial Tip?   23
        2.4.3   An Alternative Approach....................................   25
2.5   A Model for Filopodial Dynamics....................................   27
        2.5.1   Principal Thoughts About Strategic Choices..............   27
        2.5.2   Modeling F-Actin Polymerization and Diffusion   28
        2.5.3   Modeling Cytosolic Flow as a Means to Drive
                Passive Transport of G-Actin   29
        2.5.4   Model Validation   31
2.6   Conclusions...   32
References...   33

3   Fault Diagnosis with Missing Data Based on Hopfield
    Neural Networks..   37
    Raquelita Torres Cabeza, Egly Barrero Viciedo, Alberto
    Prieto-Moreno, and Valery Moreno Vega
    3.1   Introduction..   37
    3.2   Materials and Methods...   38
            3.2.1   Hopfield and Probabilistic Artificial Neural
                    Networks Models   38
            3.2.2   DAMADICS Benchmarking and Design of
                    Architectures...   41
            3.2.3   Design of the Experiments   44
    3.3   Results and Discussion..   44
    3.4   Conclusions...   45
    References...   45

4   Diagnosing Time-Dependent Incipient Faults   47
    Lídice Camps-Echevarría, Orestes Llanes-Santiago, Haroldo
    Fraga de Campos Velho, and Antônio José da Silva Neto
    4.1   Introduction..   48
    4.2   Fault Diagnosis as an Inverse Problem   49
            4.2.1   Structural Analysis   50
    4.3   Differential Evolution and Differential Evolution
            with Particle Collision   51
            4.3.1   Differential Evolution   51
            4.3.2   Differential Evolution with Particle Collision.............   52
    4.4   Two Tanks System   53
            4.4.1   Structural Analysis   55
    4.5   Experimental Methodology   56
    4.6   Results   57
    4.7   Conclusions...   60
    References...   61

**5   An Indirect Kernel Optimization Approach to Fault
      Detection with KPCA** 63
      José M. Bernal de Lázaro, Orestes Llanes-Santiago, Alberto
      Prieto-Moreno, and Diego Campos Knupp
      5.1   Introduction... 63
      5.2   Materials and Methods.................................... 65
             5.2.1   Kernel Principal Component Analysis 65
             5.2.2   Indirect Kernel Optimization Criteria 67
      5.3   Optimization with Differential Evolution 69
      5.4   Results and Discussion.................................... 70
             5.4.1   Study Case: Tennessee Eastman Process 70
             5.4.2   Experimental Description 72
             5.4.3   Discussion 72
      5.5   Conclusions... 74
      5.6   Future Work 74
      References.. 74

**6   Uncertainty Quantification in Chromatography Process
      Identification Based on Markov Chain Monte Carlo** 77
      Mirtha Irizar Mesa, Leôncio D. Tavares Câmara,
      Diego Campos-Knupp, and Antônio José da Silva Neto
      6.1   Introduction... 77
      6.2   Front Velocity Chromatography Model 78
      6.3   Parameter Uncertainty Quantification 79
             6.3.1   Delayed Rejection Adaptive Metropolis Algorithm 80
      6.4   Convergence Diagnostics 82
      6.5   Conclusions... 87
      References.. 87

**7   Inverse Analysis of a New Anomalous Diffusion Model
      Employing Maximum Likelihood and Bayesian Estimation** 89
      Diego Campos-Knupp, Luciano G. da Silva, Luiz Bevilacqua,
      Augusto C.N.R. Galeão, and Antônio José da Silva Neto
      7.1   Introduction... 90
      7.2   Direct Problem Formulation and Solution 92
      7.3   Inverse Problem 94
             7.3.1   Maximum Likelihood 94
             7.3.2   Bayesian Inference 95
      7.4   Results and Discussion.................................... 98
      7.5   Conclusions... 103
      References.. 103

**8   Accelerated Direct-Problem Solution: A Complementary
      Method for Computational Time Reduction** 105
Alberto Prieto-Moreno, Leôncio D. Tavares Câmara,
and Orestes Llanes-Santiago
  8.1   Introduction 106
  8.2   Materials and Methods 107
         8.2.1   Front Velocity 107
         8.2.2   Direct Solution 108
         8.2.3   Parameters Estimation 111
  8.3   Accelerated Simulation 112
         8.3.1   Procedure for Selecting the Solution Step 112
         8.3.2   Statistical Evaluation 113
  8.4   Results and Discussions 114
         8.4.1   Direct Solution 114
         8.4.2   Parameters Estimation 115
  8.5   Conclusions 119
  References 119

**9   Effects of Antennas on Structural Behavior
      of Telecommunication Towers** 121
Patricia Martín Rodríguez, Vivian B. Elena Parnás,
and Angel Emilio Castañeda Hevia
  9.1   Introduction 121
  9.2   Methodology 122
         9.2.1   Wind Tunnel Experiment 122
         9.2.2   Numerical Experiment 124
         9.2.3   Structure and Wind Load Modeling 126
         9.2.4   Modal Analysis 127
  9.3   Discussion of Results 128
         9.3.1   Factorial Design of Experiments with UHF
                  and VHF Antennas 128
         9.3.2   Factorial Design of Experiments Without
                  UHF and VHF Antennas 131
  9.4   Conclusions 131
  References 139

**10   Comparing Two-Level Preconditioners for Solving
      Petroleum Reservoir Simulation Problems** 141
José R.P. Rodrigues, Paulo Goldfeld, and Luiz M. Carvalho
  10.1   Introduction 141
  10.2   Methods 144
         10.2.1   ILU(k)-Based Domain Decomposition 144
         10.2.2   Coarse Space Correction 147

         10.3   Analysis and Discussion 148
                 10.3.1   Tests Description 148
                 10.3.2   Experiments... 149
                 10.3.3   Discussion 150
         10.4   Future Work 151
         References.. 151

11   **Assessment of the Reliability of Electrical Power Systems**............. 153
     Yorlandys Salgado Duarte and Alfredo M. del Castillo Serpa
         11.1   Introduction.. 153
         11.2   Materials and Methods.. 154
                 11.2.1   Models and Methods for the Adequacy Assessment...... 154
                 11.2.2   Models and Methods for the Security Assessment........ 158
                 11.2.3   Methodology... 164
         11.3   Result Analysis and Discussion 165
                 11.3.1   Analysis of the Adequacy of the RBTS 166
                 11.3.2   Analysis of the Security of RBTS 167
         11.4   Conclusions and Recommendations................................ 169
         References.. 170

12   **Polymeric Thin Film Transistors Modeling in the Presence
     of Non-Ohmic Contacts** 171
     Magali Estrada del Cueto, Antonio Cerdeira Altuzarra,
     Benjamín Iñiguez Nicolau, Lluis F. Marsal Garvi, and Josep
     Pallarés Marzal
         12.1   Introduction.. 172
         12.2   Basic Aspects of UMEM 173
                 12.2.1   Basic Expressions to Represent the Above
                          Threshold Operating Region 173
                 12.2.2   Basic Expressions to Represent the
                          Subthreshold Operating Region 175
                 12.2.3   Modeling the Output Characteristics in TFTs
                          with Non-Ohmic Contacts at D and S 176
                 12.2.4   Summarizing the Steps to Extract Model
                          Parameters when Non-Ohmic Contacts Are Present...... 177
         12.3   Examples of OTFT Characteristics Modeling 177
         12.4   Conclusions.. 177
         References.. 179

List of Contributors

Guillermo Vilalta Alonso Thermal Sciences and Fluid Department, University of São João del Rei (UFSJ), São João del Rei, Brazil

José A. Vilalta Alonso Industrial Engineering Department, Instituto Superior Politécnico José Antonio Echeverría (CUJAE), Havana, Cuba

Antonio Cerdeira Altuzarra Electrical Engineering Department, CINVESTAV-IPN, México, México

Luiz Bevilacqua Universidade Federal do Rio de Janeiro, COPPE-UFRJ, Rio de Janeiro, Brazil

Raquelita Torres Cabeza Automatic and Computing Department, Instituto Superior Politécnico José Antonio Echeverría (CUJAE), Marianao, Cuba

Leôncio D. Tavares Câmara Mechanical Engineering and Energy Department, IPRJ-UERJ, Nova Friburgo, Brazil

Diego Campos-Knupp Mechanical Engineering and Energy Department, Polytechnic Institute, IPRJ-UERJ, Nova Friburgo, Brazil

Lídice Camps-Echevarría CEMAT, Instituto Superior Politécnico José Antonio Echeverría (CUJAE), Marianao, Cuba

Luiz M. Carvalho Departamento de Matemática Aplicada, Universidade do Estado do Rio de Janeiro (UERJ), Rio de Janeiro, Brazil

Luciano G. da Silva Mechanical Engineering and Energy Department, IPRJ-UERJ, Nova Friburgo, Brazil

Haroldo Fraga de Campos Velho INPE, São José dos Campos, Brazil

José M. Bernal de Lázaro Reference Center for Advanced Education, Instituto Superior Politécnico José Antonio Echeverría (CUJAE), Marianao, Cuba

Carlos A. de Moura Instituto de Matemática e Estatística - IME, Universidade do Estado do Rio de Janeiro - UERJ, Rio de Janeiro, Brazil

Alfredo M. del Castillo Serpa CEMAT, Instituto Superior Politécnico José Antonio Echeverría (CUJAE), Marianao, Cuba

Magali Estrada del Cueto Electrical Engineering Department, CINVESTAV-IPN, México, México

Yorlandys Salgado Duarte CEMAT, Instituto Superior Politécnico José Antonio Echeverría (CUJAE), Marianao, Cuba

Augusto C.N.R. Galeão Laboratório Nacional de Computação Científica, LNCC, Petrópolis, Brazil

Lluis F. Marsal Garvi Department of Electronic, Electric and Automation Engineering, Universitat Rovira i Virgili, Tarragona, Spain

Paulo Goldfeld Instituto de Matemática, Universidade Federal do Rio de Janeiro, Rio de Janeiro, Brazil

Angel Emilio Castañeda Hevia Facultad de Ingeniería Civil, Instituto Superior Politécnico José Antonio Echeverrja (CUJAE), Marianao, Cuba

Mauricio V. Kritz Laboratório Nacional de Computação Científica, LNCC, Petrópolis, Brazil

Thiago F. Leal Graduate Program on Mechanical Engineering, Universidade do Estado do Rio de Janeiro - UERJ, Rio de Janeiro, Brazil
Instituto Federal do Rio de Janeiro - IFRJ, Paracambi, Brazil

Laurentiu Lipsa CARTIF Centro Tecnológico, Boecillo, Spain

Orestes Llanes-Santiago Automatic and Computing Department, Instituto Superior Politécnico José Antonio Echeverría (CUJAE), Marianao, Cuba

Josep Pallarés Marzal Department of Electronic, Electric and Automation Engineering, Universitat Rovira i Virgili, Tarragona, Spain

Mirtha Irizar Mesa Automatic and Computing Department, Instituto Superior Politécnico José Antonio Echeverría (CUJAE), Marianao, Cuba

Antônio J. Silva Neto Mechanical Engineering and Energy Department, IPRJ-UERJ, Nova Friburgo, Brazil

Benjamín Iñiguez Nicolau Department of Electronic, Electric and Automation Engineering, Universitat Rovira i Virgili, Tarragona, Spain

Félix Nieto CARTIF Centro Tecnológico, Boecillo, Spain

Vivian B. Elena Parnás Facultad de Ingeniería Civil, Instituto Superior Politécnico José Antonio Echeverrja (CUJAE), Marianao, Cuba

María Ángeles Pérez Institute of Advanced Production Technologies (ITAP), University of Valladolid, Valladolid, Spain

Alberto Prieto-Moreno Automatic and Computing Department, Instituto Superior Politécnico José Antonio Echeverría (CUJAE), Marianao, Cuba

Andreas Prokop Faculty of Life Sciences—FLS, University of Manchester, Manchester, UK

José R.P. Rodrigues Petrobras/CENPES, Rio de Janeiro, Brazil

Patricia Martín Rodríguez Facultad de Ingeniería Civil, Instituto Superior Politécnico José Antonio Echeverrja (CUJAE), Marianao, Cuba

Eduardo Soudah International Center for Numerical Methods in Engineering (CIMNE), Barcelona, Spain

Carlos Vaquero University and Clinic Hospital of Valladolid, Valladolid, Spain

Valery Moreno Vega Automatic and Computing Department, Instituto Superior Politécnico José Antonio Echeverría (CUJAE), Marianao, Cuba

Egly Barrero Viciedo Automatic and Computing Department, Instituto Superior Politécnico José Antonio Echeverría (CUJAE), Marianao, Cuba

Acronyms

AAA	Abdominal aortic aneurysm
ANN	Artificial neural network
AUC	Area under the ROC curve
CAPES	Coordenação de Aperfeiçoamento de Pessoal de Nível Superior
CFD	Computational fluid dynamics
CNPq	Conselho Nacional de Desenvolvimento Científico e Tecnológico
CT	Ultrasound or computed tomography
DAMADICS	Development and Application of Methods for Actuator Diagnosis in Industrial Control Systems
DE	Differential evolution
DEwPC	Differential evolution with particle collision
DRAM	Delayed rejection adaptive Metropolis algorithm
EENS	Expected energy not supplied
EPS	Electrical power system
F	Scale factor
FAPERJ	Fundação Carlos Chagas Filho de Amparo à Pesquisa do Estado do Rio de Janeiro
FAR	False alarm rate
FDI	Fault diagnosis
FDR	False detection rate
FEM	Finite element method
FOR	Forced outage rate
FSI	Fluid-solid interaction
GMRES	Generalized minimum residual method
GPU	Graphics processing unit
HL I	Hierarchical level I
HL II	Hierarchical level II
HOMO	Highest occupied molecular orbital
IEEE RTS	IEEE Reliability Test System
IP	Incomplete product
KPCA	Kernel Principal Component Analysis

IFS	Incomplete forward substitution
ILU	Incomplete LU
ILU(k)	Level-k ILU
ILT	Intraluminal thrombus
LDC	Load duration curve
LOLE	Loss of load expectation
LOLP	Loss of load probability
LU	Lower upper factorization
MAP	Maximum a posteriori
MATLAB	Matrix Laboratory
MCMC	Markov chain Monte Carlo
MECORE	Monte Carlo Simulation and Enumeration Composite System Reliability Evaluation Program
MES	Higher Education Ministry of Cuba
NNZ	Number of nonzeros
NP	Population size
OTFTs	Organic thin film transistors
P3HT	Poly(3-hexylthiophene)
PCA	Particle collision algorithm
PCs	Principal components
PDE	Partial differential equation
PETSc	Portable, extensible toolkit for scientific computation
PIP	Peak intraluminal pressure
PMMA	Polymethyl metacrylate
PNN	Probabilistic neural network
PTFTs	Polymeric thin film transistors
PWSS	Peak wall shear stress
RAM	Random access memory
RAS	Restricted additive Schwarz
RBF	Radial basis function
RBTS	Roy Billinton Test System
RPI	Rensselaer Polytechnic Institute
SA	Simulated annealing
SCADA	Supervisory and Data Acquisition System
SIMULINK	Simulation and Model-Based Design
SMB	Simulated moving bed
SMCNS	Simulation of Monte Carlo nonsequential
SMCS	Simulation of Monte Carlo sequential
SPE	The Society of Petroleum Engineers
TE	Tennessee Eastman process
TFTs	Thin film transistors
UHF	Ultra high frequency
UMEM	Unified model and parameter extraction method
UOTFT	Universal Organic TFT Model for Circuit Design

VHF	Very high frequency
VMTK	Vascular Modeling ToolKit
WAS	Weighted additive Schwarz
WSS	Wall shear stress

Chapter 1
Preliminary Correlations for Characterizing the Morphology of Abdominal Aortic Aneurysms as Predictor of Rupture

Guillermo Vilalta Alonso, Eduardo Soudah, José A. Vilalta Alonso, Laurentiu Lipsa, Félix Nieto, María Ángeles Pérez, and Carlos Vaquero

Abstract The morphology of abdominal aortic aneurysms (AAA) has been recognized as a factor that may predispose their rupture. The time variation of the AAA morphology induces hemodynamic changes in morphological behavior that, in turn, alters the distribution of hemodynamic stress on the arterial wall. This behavior can influence the phenomenon of rupture. In order to evaluate the relationship between the main geometric parameters characterizing the AAA and the hemodynamic stresses, 6 AAA models were reconstructed and characterized. The models were characterized using thirteen geometrical factors based on the lumen center line: eight 1D indices, three 3D indices, and two 0D indices. The temporal and spatial distributions of hemodynamic stresses were computed using computational

G.V. Alonso (✉)
Thermal Sciences and Fluid Department, University of São João del Rei (UFSJ),
São João del Rei, 36307-352, Brazil
e-mail: gvilalta@ufsj.edu.br

E. Soudah
International Center for Numerical Methods in Engineering (CIMNE), Barcelona, Spain
e-mail: esoudah@cimne.upc.edu

J.A.V. Alonso
Industrial Engineering Department, Instituto Superior Politécnico José Antonio Echeverría
(CUJAE), Havana, CP 19390, Cuba
e-mail: jvilalta@ind.cujae.edu.cu

L. Lipsa • F. Nieto
CARTIF Centro Tecnológico, Boecillo, 47151, Spain
e-mail: liplau@cartif.es; feline@cartif.es

M.A. Pérez
Institute of Advanced Production Technologies (ITAP), University of Valladolid, Valladolid,
47011, Spain
e-mail: marrue@cartif.es

C. Vaquero
University and Clinic Hospital of Valladolid, Valladolid, 47011, Spain
e-mail: cvaquero@med.uva.es

© Springer International Publishing Switzerland 2016
A.J. da Silva Neto et al. (eds.), *Mathematical Modeling and Computational Intelligence in Engineering Applications*, DOI 10.1007/978-3-319-38869-4_1

fluid dynamics. The results showed that the hemodynamic stresses are modified by the time variations of the AAA morphology, and therefore, the hemodynamic stresses, in combination with other parameters, could be a criterion for improved rupture risk prediction. Statistical correlations between hemodynamic stresses and geometric indices have confirmed the influence by the AAA morphometry on the prediction of the rupture risks, although higher reliability of these correlations is required.

Keywords Abdominal Aortic Aneurysm (AAA) • Mathematical modeling • Morphology • Hemodynamic Stresses • Pearson coefficient • Computational Fluid Dynamics (CFD) • Rupture risk • Prediction • Lumen center line

1.1 Introduction

The morphology of the abdominal aortic aneurysms (AAA) has been recognized as a factor that may predispose their rupture. This aspect is at the core of the current clinic treatment of this pathology: the risk of rupture is essentially assessed on the basis of its maximum transverse diameter. Depending on the size of the aneurysm, as defined by its maximum transverse diameter at the moment of its diagnosis, the patient either undergoes a process of repair to avoid the rupture or is kept under observation. Recent results [10] have corroborated that, even if validated by a significant empirical data, this criterion is not sufficiently precise to support a personalized assessment of rupture risk, and it often leads to fail. Several clinical studies have shown that the rupture risk of an AAA with diameter smaller than 50 mm is within the 12.8–23 % range [4]. However, in other cases [5], where diameters are much higher than the threshold value (55–60 mm), the rupture has not occurred. Given these considerations, these morphological parameters cannot be considered as the only criterion. It is widely accepted nowadays that bio-mechanical and biological factors can influence the risk of rupture. The temporal variation of the morphology of the AAA determines modifications of the hemodynamic behavior, that, in turn, alters the spatial and temporal distribution of the hemodynamic stresses on the arterial surface; and as a result, a two-way relationship is established, which can modify the rupture phenomenon. Recent studies [3, 9] have confirmed that the maximum value of wall stress in case of aneurysm is a more reliable predictor of rupture of AAA, as compared with the maximum transverse diameter. The peak wall stress is associated with the geometry of the aneurysm [8, 11], which reveals the importance of the morphometric characterization of the AAA. The need for the quantification of the morphometry of the AAA in each case under examination is, therefore, evident; and this quantification can constitute a valid method to make personalized assessment of the risk of rupture[7].

Definitely, the wall stress, by itself, is not a sufficient factor to predict the risk of rupture. Consideration must also be given to the local estimation of the arterial wall strength. From a purely mechanical perspective, the AAA rupture occurs when the stress on the internal wall of the aneurysm overcomes the capacity of the arterial tissue to resist to this stress. The purpose of this chapter is to determine preliminary correlations between the hemodynamic stress and the morphometry of the AAA, as a way to improve the risk of rupture prediction.

1.2 Material and Methods

1.2.1 AAA Geometry: Segmentation and Reconstruction

The aneurysms selected for this study are from patients being treated at the Valladolid Clinical University Hospital, Spain. In all cases, the patients gave their consent for the use of their data in this study that has been endorsed by the Ethical Committees of the participating institutions. The lumen of AAA was segmented using MeVisLab and Vmtk; i.e., image processing software packages. A semi-automatic method was applied to obtain properly defined geometries.

After the lumen was segmented, the 3D surfaces were automatically generated, and finally, the geometry of the 3D volumes was obtained. Given the poor quality elements in the final surface, the rise of sharply edged peaks, and the presence of stairs effect, it was necessary that the surface be optimized by improving its grid and applying smoothing techniques. The algorithm applied in this case does not cause the surface morphology to become reduced or distorted, since low-resolution filters were used, and these filters only affect the elements with high level of curvature.

1.2.2 AAA Geometry: Size and Shape Indices

Recent studies have focused on the definition of indices that help geometrically characterize AAA based on lumen center line. For the purpose of the AAA geometric characterization, thirteen size and shape indices were defined by means of a user-defined algorithm based on lumen center line. In this chapter, eight 1D indices were determined and applied as follows: maximum transversal diameter D, aneurismal length L, proximal neck diameter D_{pn}, distal neck diameter D_{dn}, left iliac diameter D_{li}, right iliac diameter D_{ri}, bifurcation angle α, and asymmetric ϵ. All these indices are directly measured from CT. The 3D indices that were determined and applied are: tortuosity T, curvature C, and aneurysmal sac volume V_{AAA}. The 0D indices are defined by means of appropriate relationship between these 1D indices: saccular index (relationship between D and L) and deformation rate γ (relationship between D_{pn} and D). Figure 1.1 shows a schematic representation of the AAA, indicating the geometric indices utilized herein.

Using the procedure described above, six AAAs were selected and segmented. As shown in Fig. 1.2, all the aneurysms have a geometry that is characterized by their asymmetry, superficial complexity, and their twisting degree of a segment near the iliac arteries. These features are expected to exert a strong influence on the flow field within the aneurysmatic sac. The information for the 3D reconstruction of AAA

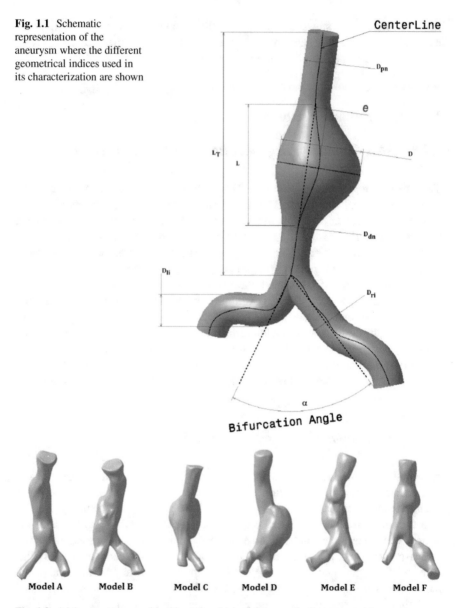

Fig. 1.1 Schematic representation of the aneurysm where the different geometrical indices used in its characterization are shown

Fig. 1.2 AAA geometries used in this study, obtained from medically processed images

was directly extracted from a high-resolution computerized tomography scan, with contrast increase. Figure 1.2 shows the six segmented and reconstructed aneurysm models used in this study.

1.2.3 Computational Methods

CFD analysis was conducted using BioDyn; i.e., a user-friendly interface based on the commercially available software package Tdyn [2]. This package is a fluid dynamics and multi-physics based simulation environment that relies on the stabilized finite element method used to solve the Navier–Stokes equations. Blood was characterized as a Newtonian, homogeneous, and incompressible fluid, with constant $\rho = 1040\,\mathrm{Kg/m^3}$ and dynamic viscosity μ=0.004 Pa s. Mathematically, its boundary conditions can be expressed as:

$$V = 0|_{wall} \tag{1.1}$$

$$u_z = 2 * u(t) * (1 - 2r/dr)^2; u_r|_{r=0} = 0 \tag{1.2}$$

$$\tau = \hat{n} * p(t) * I * \hat{n} \tag{1.3}$$

A no-slip condition (rigid vessel wall) was imposed on the surface of the AAA, as defined by (1.1). This choice was attributed to the fact that the physiological parameters characterizing the arterial mechanical behavior of the AAA wall were not well determined. This approach, however, considerably reduces the discretization effort; in particular, the boundary-layer gridding as well as the computational cost. Notwithstanding, other approaches have applied fluid structure interaction models [4].

Inlet velocity waveforms were obtained from quantitative measurements in the abdominal region. A transient blood flow was imposed on the abdominal aorta (approximately above the infra renal arteries) [6]. Velocity U was calculated for each patient in order to obtain a total fluid volumetric flow rate of 350 ml for an entire cardiac cycle. The outlet boundaries were located at the common iliac arteries, where the pressure follows pulsatile waveforms. The inlet velocity is assumed as a fully developed parabolic profile at the inlet (see (1.2)), where dr is the inner radius of the abdominal aorta, ur is the Cartesian components of the velocity vector in the z direction, $u(t)$ and $p(t)$ are the time-dependent velocity and pressure waveforms used [6]. The pressure boundary conditions are given by (1.3), where τ_n is the normal traction at the outlet; I is the standard identity matrix, and $\hat{n}$ designates the normal of the respective boundary.

Figure 1.3 shows the two above-mentioned boundary conditions. The left image is the inlet velocity waveform profile and the right image is the outlet pressure profile.

The simulation involved 50 initial steps to stabilize the initial condition solution. The time integration method chosen was a Backward Euler, using a biconjugate gradient non-symmetric solver designed to accelerate the calculation time performance. The procedure also covered 4th order pressure stabilization and automatic velocity advection stabilization [1]. Three cardiac cycles were run for the purpose of obtaining a stationary result. For each AAA, the wall shear stress (WSS) distribution, velocity, vorticity, and maximum pressure distribution over the aneurysmal *sac* were calculated.

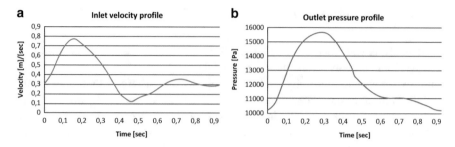

Fig. 1.3 Boundary conditions for the hemodynamic simulations

1.2.4 Statistical Analysis

The Pearson correlation coefficient was used to assess the linear relation between the main hemodynamic parameters (peak wall shear stress (PWSS) and maximum intraluminal pressure) and the predefined morphological indices for the purpose of evaluating the AAA rupture risk. The decision to use the aforementioned coefficient was based on the advantage posed by metric scale of the considered variables. Tested hypotheses also supported the statistical significance of the correlations ($p \leq 0,15$ was considered significant). The PWSS values ranged between 0,414 and 17,60 Pa and its mean value was 7,15 Pa. The maximum intraluminal pressure values (PIP) ranged between 10,188 and 16,565 Pa and its mean value was 16,185 Pa. A software package, Minitab for Windows, release 15.0, standard version was used to obtain these results.

1.3 Results and Discussions

According to the dominant ethological hypothesis, the AAAs are generated by the interconnection between the structural changes of the internal components of the arterial wall (at molecular level) and the altered flows patterns of hemodynamic stress that acts on the blood-vessel walls. Any structural change in the surface can affect the flow pattern through the abdominal aorta.

Similarly, any change in the blood flow patterns can alter the wall stress distribution and the hemodynamic pressure, thus giving rise to several disruptive processes, including but not limited to the calcification and inflammation of the wall [2]. Previous studies have shown that a perturbed flow, as is the usual case with AAAs, can help cause vascular diseases, mainly through its harmful effects on the endothelial tissue.

Subject to the AAA shape as well as the pressure and velocity waves described as baseline conditions, when the pressure rises to its top systolic value, the blood flow velocity is in its last systolic phase; and in this condition, the blood flow through the neck starts to slow. As a consequence, recirculating regions are developed, and these regions determine a locally and temporarily stabilized distribution of the hemodynamic stress. The behavior of the hemodynamic variables, therefore, is strongly affected by the behavior of the flow that, in this case, is characterized by regular and temporal acceleration and deceleration, depending on the geometry of the AAA. In any phase of the cycle, the pressure distribution remains almost constant on the wall of the aneurysm. The results show that the high complexity of the surface around the aneurysmal sac, characterized by a divergent–convergent shape, apparently has little effect on the distribution of the surface, except for the times when a change of phase in the wave of the flow occurs.

Figure 1.4 (right) shows the superficial distribution of the absolute maximum pressure in three cases: A, C, and D. As can be observed, the distribution of the inter-luminal pressure is highly correlated with the geometric characterization of the aneurysm. In case A, where the maximum diameter is quite small and the length is relatively large, the pressure value in the internal surface is smaller than the value recorded for the other two cases. The changes in the wall pressure distribution are also time-dependent, with similar patterns at different points of the surface, and this distribution tends to follow the pulse wave. The highest surface pressure gradients are recorded during systolic acceleration. For this reason, the flow remains laminar without separation from the surface of the aneurysmal sac, and the highest adverse surface pressure gradients are recorded during systolic deceleration, occurring at the start of the inverse flow period. An interesting behavior is observed near the distal neck, where an increase in the characteristic surface pressure values can be seen in the flow deceleration phase; alternating with an abrupt pressure fall during the pulse-flow acceleration phase. The results also show that the walls of the aneurysm are under low intensity stress during the cardiac cycle, except for their proximal and distal zones, where the tangential surface stress registers high values during the systolic phase. These results are consistent with the values registered in the flow pattern of the same regions. The changes in the surface stress distribution are likely time-dependent. In general, the behavior of this variable can (1) determine degenerative lesions in the walls of the aneurysm; (2) alter the thickness and mechanical characteristics blood-vessel walls; and (3) cause the risk of rupture to increase. The behavior of the distribution of the absolute maximum surface stress (defined as the top stress value recorded for each point of the geometric dominion in all the temporal dominions) is shown in Fig. 1.4 (left), for cases A, C, and D.

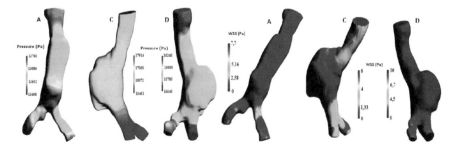

Fig. 1.4 AAA geometries used in this study, as obtained from medically processed images

The above-described behavior has been associated with the process of intraluminal thrombus (ILT) formation. In approximately 70 % of the total aneurysms, ILT has been found. A possible hypothesis is that the formation of the ILT may be the result from a continuous process characterized by activated platelets introduced through the blood flow and settled in non-endothelialized regions (endothelially dysfunctional regions) of the surface; that is, the regions exposed to low tangential stress (WSS).

The flow pattern in the internal part of the aneurysmatic sac shows that the flow field is dominated by vortices in the regions close to the wall of the aneurysmatic sac, where basically sharp curvatures exist.

The typical behavior of the flow field for the six case studies can be summarized as follows: The beginning of each cardiac cycle is characterized by residual vortices from the preceding cycle. The progressive acceleration of the flow determines an increase in the number of these residual vortices that manage to occupy a significant part of the aneurysm sac at its rear wall and generate a stream flow close to its frontal wall. As the maximum systolic value nears, the vortices are ejected at the inlet of the aneurysmatic sac. The maximum velocity values, as well as the velocity gradients for the dominion, are obtained during maximum systolic pressure. The temporal deceleration, combined with rising pressure, causes weak convective effects, and as a consequence, the fast stream flow gives rise to significant disturbance in the flow field.

The reversion flow that occurs approximately in the middle of the cycle causes the recirculation flow intensity to decrease and the vortices to move close to the center, in the direction of the neck. During the last diastolic phase, the flow comes back to its main direction; and as a result, the main vortices shift closely to the distal neck. The final part of this phase is characterized by an almost constant flow and an intensification of the perturbations, due to local velocity increases.

1.3.1 Relationship Between Hemodynamic Parameters and Geometric Indices of the AAA

A total of six AAA models were simulated for this study. The assessment of the correlations between the main hemodynamic parameters and the different indices that characterized the AAA from a geometric perspective was an important component of this effort. The correlation study and its results, as presented here, are based on the calculation of the Pearson correlation coefficients (r) and the execution of its corresponding test of hypothesis ($H_0 : \rho = 0$) to determine their statistical significance. The p values were taken into account and scatterplots were analyzed in the pursuit of more accurate correlations to assess the risk of rupture of AAA. The most important results are presented and discussed below.

1.3.1.1 Peak Wall Shear Stress

PWSS vs. 1D Indices The relationships between PWSS and the different 1D indices defined in this study are illustrated in Fig. 1.5: (a) PWSS vs. D, (b) PWSS vs. L, (c) PWSS vs. D_{pn}, (d) PWSS vs. D_{dn}, (e) PWSS vs. D_{li}, (f) PWSS vs. D_{ri}, (g) PWSS vs. bifurcation angle, and (h) PWSS vs. Asymmetry. A correlation assessment of the PWSS with these indices resulted in the following statistical parameter: bifurcation angle ($r = 0.664, p = 0.141$) which is significant at 15 %. The other geometric parameters do not correlate significantly with PWSS.

PWSS vs. 3D Indices The relationships between PWSS and the different 3D indices defined in this study are illustrated in Fig. 1.6: (a) PWSS vs. Curvature (C), (b) PWSS vs. Tortuosity (T), (c) PWSS vs. V_{AAA}. These geometric parameters do not correlate significantly with PWSS.

PWSS vs. 0D Indices The relationships between PWSS and the different 0D indices defined in this study are illustrated in Fig. 1.7: (a) PWSS vs. γ, (b) PWSS vs. χ. These geometric parameters do not correlate significantly with PWSS.

In a recent paper [8] using 28 virtual AAA, its authors showed the existence of significant correlation between PWSS, on the one hand, and the maximum diameter D, asymmetry e, and deformation rate χ individually, on the other. The mismatch correlation results between PWSS and L are attributed to the assumption of L values that were higher than the actual values obtained from medically processed images, as discussed above.

1.3.1.2 Peak Intraluminal Pressure

PIP vs. 1D Indices The relationships between PIP and the different 1D indices defined in this study are illustrated in Fig. 1.8: (a) PIP vs. D, (b) PIP vs. L, (c) PIP vs. D_{pn}, (d) PIP vs. D_{dn}, (e) PIP vs. D_{li}, (f) PIP vs. D_{ri}, (g) PIP vs. bifurcation angle, and (h) PIP vs. Asymmetry. The PIP was best correlated with D_{dn} ($r = -0.656, p = 0.15$). The rest of the geometric parameters do not correlate significantly with PIP.

PIP vs. 3D Indices The relationships between PIP and the different 3D indices defined in this study are illustrated in Fig. 1.9: (a) PIP vs. C, (b) PIP vs. T, (c) PIP vs. V_{AAA}. A significant negative correlation ($r = -0.661, p = 0.143$) was observed between PIP and V_{AAA}.

PIP vs. 0D Indices The relationships between PIP and the different 0D indices defined in this study are illustrated in Fig. 1.10: (a) PIP vs. γ, (b) PIP vs. χ. These geometric parameters do not correlate significantly with PIP.

These results are consistent with other studies in which the PIP has been established as a dominant factor when the risk of rupture is assessed. The main limitation of the results herein is their reliance on a small data sampling size: six cases. The research team that produced this study is currently working on two approaches to improve the reliability of rupture risk prediction. The first

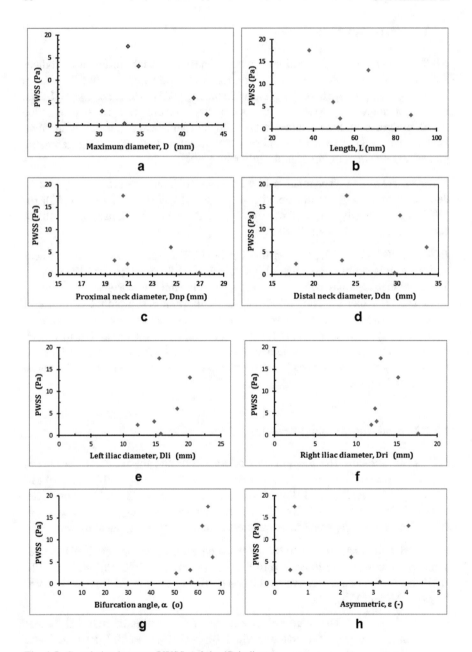

Fig. 1.5 Correlation between PWSS and the 1D indices

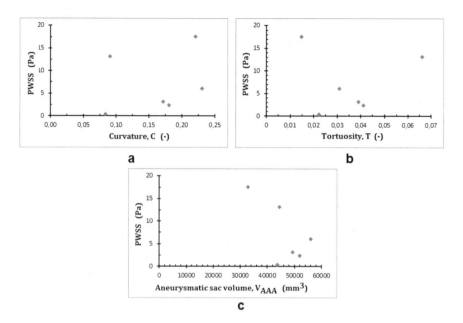

Fig. 1.6 Correlation between PWSS and the 3D indices

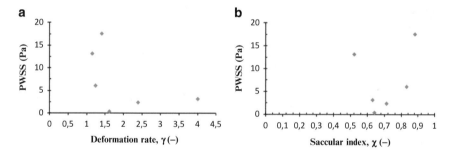

Fig. 1.7 Correlation between PWSS and the 0D indices

approach calls for an expanded analysis (segmentation, geometric characterization of the AAA, and computer simulation) to include a larger number of cases. The second approach involves the application of statistical methods based on resampling techniques (bootstrap) to overcome the effect of the small sample size.

1.4 Conclusions

This paper discusses a numerical study of six patient-specific AAAs designed to obtain information that help better understand the relationships between the hemodynamic stresses and geometrical index associated with risk of rupture of

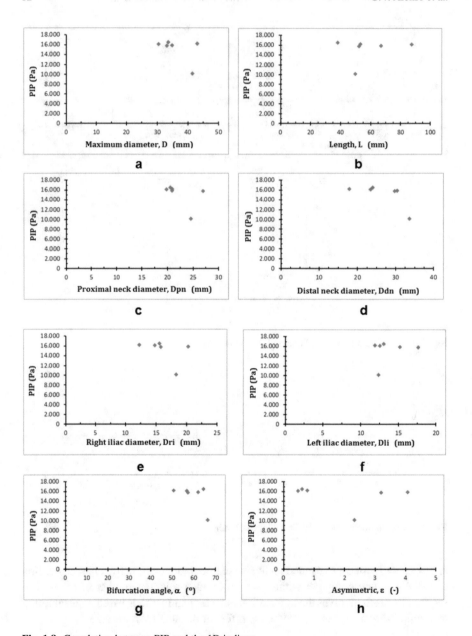

Fig. 1.8 Correlation between PIP and the 1D indices

AAA. Its main findings are as follows: the AAA rupture risk could be better predicted using a morphological index, combined with temporal and regional distribution of hemodynamic stresses. The flow field pattern within aneurysmal sac is characterized by vortices that can activate platelets, which in turn trigger the

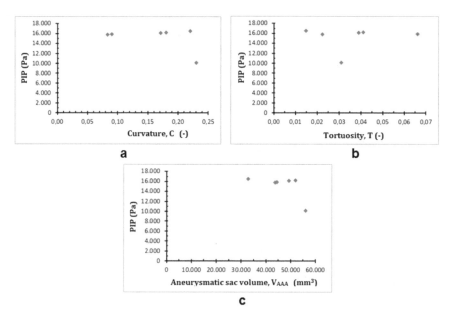

Fig. 1.9 Correlation between PIP and the 3D indices

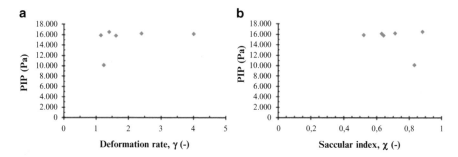

Fig. 1.10 Correlation between PIP and the 0D indices

formation of ILT. These intraluminal thrombi have been found to be a potential source for increased rupture risk. For the six individual cases covered by this study, PWSS was correlated with the size and shape indices defined herein. Among the 1D indices, the bifurcation angle was significant at 15 %. For 3D index, aneurysmal sac volume had the same level of significance. For the six participating cases, the PIP was correlated with the size and shape indices defined herein. Among the 1D indices, distal neck diameter was significant at 15 %. For the 3D index, aneurysmal sac volume had the same level of significance. For the purpose of generating more reliable data, the sampling size is currently being expanded, and other statistical tools, such as Bootstrap, will be applied to the determination of correlations between morphological index and hemodynamic stresses.

Acknowledgements The authors are grateful to the Fundação de Amparo à Pesquisa do Estado de Minas Gerais, (FAPEMIG)/Brazil, for its financial support for the publication of their research results.

References

1. Barrett, R., Berry, M., Chan, T., Demmel, J., Donato, J., Dongarra, J., Eijkhout, V., Pozo, R., Romine, C., der Vorst., H.V.: Templates for the Solution of Linear Systems: Building Blocks for Iterative Methods, 2 edn. SIAM, Philadelphia (1994)
2. Bordone, M.: Biodyn user manual. Tdyn: theoretical background. Technical Report, Compass (2012). Http://www.compassis.com/compass
3. Fillinger, M., Marra, S., Raghavan, M., Kennedy, F.: Prediction of rupture risk in abdominal aortic aneurysm during observation: wall stress versus diameter. J. Vasc. Surg. **37**, 724–732 (2003)
4. Leung, J., Wright, A., Cheshire, K., Crane, J., Thom, S., Hughes, A., Xu, Y.: Fluid structure interaction of patient specific abdominal aortic aneurysm: a comparison with solid stress model. Biomed. Eng. Online **5**(33), (2006). doi: 10.1186/1475-925X-5-33
5. Papaharilaou, Y., Ekaterinaris, J., Manousaki, E., Katsamouris, A.: A decoupled fluid structure approach for estimating wall stress in abdominal aortic aneurysm. J. Biomech. **40**(2), 367–377 (2007)
6. Pedersen, E., Kozerke, S., Ringgaard, S., Scheidegger, M., Boesiger, P.: Quantitative abdominal aortic flow measurements at controlled levels of ergometer exercise. Magn. Reson. Imaging **17**(4), 489–494 (1999)
7. Shum, J., Martufi, G., Dimartino, E., Washintong, C., Grisafi, J., Muluk, S., Finol, E.A.: Quantitative assessment of abdominal aortic aneurysm geometry. Ann. Biomed. Eng. **39**(1), 273–289 (2011)
8. Soudah, E., Vilalta, G., Bordone, M., Nieto, F., Vilalta, J., Vaquero, C.: Estudio paramétrico de tensiones hemodinámicas en modelos de aneurismas de aorta abdominal. Revista Internacional de Métodos Numéricos para Cálculo y Diseño en Ingeniería. **31**(2), 106–112 (2015)
9. Venkatasubramaniam, A., Fagan, M., Mehta, T., Mylankal, K., Ray, B., Kuhan, G., Chetter, I., McCollum., P.: A comparative study of aortic wall stress using finite element analysis for ruptured and non-ruptured abdominal aortic aneurysms. Eur. J. Vasc. Endovasc. Surg. **28**(2), 168–176 (2004)
10. Vorp, D.A.: Biomechanics of abdominal aortic aneurysm. J. Biomech. **40**, 1887–1902 (2007)
11. Vorp, D.A., Raghavan, M., Webster, M.: Mechanical wall stress in abdominal aortic aneurysm: influence of diameter and asymmetry. J. Vasc. Surg. **27**(4), 632–639 (1998)

Chapter 2
Mathematical-Computational Simulation of Cytoskeletal Dynamics

Carlos A. de Moura, Mauricio V. Kritz, Thiago F. Leal, and Andreas Prokop

Abstract Actin and microtubules are components of the cytoskeleton, and are key mediators of neuron growth and maintenance. Knowing how they are regulated enhances our understanding of neural development, ageing, degeneration, and regeneration. However, biological investigation alone will not unravel the complex cytoskeletal machinery. We expect that inquiries about the cytoskeleton can be significantly enhanced if their physico-chemical behavior is concealed and summarized in mathematical and computational models that can be coupled to concepts of biological regulation. Our computational modeling concerns the mechanical aspects associated with the dynamics of relatively simple, finger-like membrane protrusions called filopodia. Here we propose an alternative approach for representing the displacement of molecules and cytoplasmic fluid in the extremely narrow and long filopodia and discuss strategies to couple the particle-in-cell method with algorithms for laminar flow to model the two phases of actin dynamics: polymerization into filaments which are pulled back into the cell and compensatory G-actin drift towards its tip to supply polymerization. We use nerve cells of the fruit fly *Drosophila* as an effective, genetically amenable biological system to generate experimental data as the basis for the abstract models and their validation.

C.A. de Moura (✉)
Instituto de Matemática e Estatística - IME, Universidade do Estado
do Rio de Janeiro - UERJ, Rio de Janeiro, Brazil
e-mail: demoura@ime.uerj.br

M.V. Kritz
Laboratório Nacional de Computação Científica - LNCC, Petrópolis, RJ, Brazil
e-mail: kritz@lncc.br

T.F. Leal
Graduate Program on Mechanical Engineering, Universidade do Estado do Rio de
Janeiro - UERJ, Rio de Janeiro, Brazil

Instituto Federal do Rio de Janeiro - IFRJ, Paracambi, Brazil
e-mail: thiagofranco@ime.uerj.br

A. Prokop
Faculty of Life Sciences - FLS, University of Manchester, Manchester , UK
e-mail: Andreas.Prokop@manchester.ac.uk

© Springer International Publishing Switzerland 2016
A.J. da Silva Neto et al. (eds.), *Mathematical Modeling and Computational Intelligence in Engineering Applications*, DOI 10.1007/978-3-319-38869-4_2

Keywords Computational biology • Mathematical modeling • Particle-in-cell methods • Drosophila • Filopodia • Cytoskeleton • Actin dynamics • Cytosol flow

2.1   Introduction

Biological phenomena rely on physical and chemical interactions. As elements of a physical phenomenon, biological components interact freely with one another: whenever in contact they exchange mass and energy. Their interactions as chemical elements are not as free, since they depend on chemical compatibilities of the interacting elements. This makes biological interactions highly complex. The elements giving cause to biological phenomena are dynamic, self-regulatory, and resilient, and entities often interact through the exchange of information (signals), without being necessarily in contact or even close together.

To understand biological systems at the sub-cellular scale, we need to disentangle the organizational (functional) status of biological elements, as well as the regulatory elements, interactions, and rules underlying their dynamics, from purely biochemical reactions. All biological processes are grounded on physical and chemical interactions but some of them have only sense within cells or other organisms, having no reason to spontaneously exist elsewhere.

A way to achieve this is to have the physico-chemical dynamics well described and operationally documented as a working and consensual mathematical/computational model amenable to manipulations at higher levels, including by regulatory mechanisms induced by signaling and other stimuli. This can be achieved through the employment of mathematics and computer science to represent the behavior of complex systems occurring in life-phenomena, through structured algorithm-based simulations [6, 11, 24, 25, 37]. The key purpose is to represent or reproduce properties that the system under analysis displays.

Both types of observation, real experimental data and computer simulation, complement each other in several ways. Computational modeling depends on data, insights and concepts/cartoon models derived from real experimentation (in our case with biological systems). Vice versa, mathematical/computational models offer several advantages to advance the understanding of biological systems: (a) they offer inclusion of wider ranges of parameters far beyond the capacity of the human mind, (b) they can operate beyond space and time scales that can be experimentally addressed, (c) allow variables and geometries to be easily adjusted to study their impacts on the system, (d) make predictions or suggest the existence of particular interactions or structures that can then be tested experimentally, (e) provide opportunities to test assumptions and theoretical knowledge in order to refine hypotheses or suggest what might be true, (f) determine which experimental variables are most important in a system, and (g) can be used to synthesize experimental data and for testing data interpretation methods. The Nobel Prize winning physicist Kenneth Wilson proposed in the late 1980s that, for different sciences, computational modeling ought to gain the same status as theoretical analyses and laboratory experiments [59]. This enlarges the possibilities inherent in the scientific hypothesis-deduction-observation cycle and relaxes the constraints imposed by the impossibility of performing certain experiments.

We aim to model biological problems and, for this, we focus on the cytoskeleton. The cytoskeleton comprises mechano-resistant, yet highly dynamic, networks composed of filamentous protein polymers, called actin (thin filaments; diameter of 5–7 nm), intermediate filaments (5–12 nm), and microtubules (thick filaments; ∼25 nm), which support cell architecture and dynamics [20]. There is virtually no cell function that does not depend on the cytoskeleton, yet the number of essential proteins binding and regulating the cytoskeleton is surprisingly low [15, 49]. This is possible because the same cytoskeletal regulating proteins can be employed in different contexts, contributing to very distinct cytoskeletal networks and dynamics. Therefore, genetic defects of any of these components will have multiple (pleiotropic) effects, and the many human diseases caused by such defects [49] are by nature complex systemic phenomena. Furthermore, cytoskeletal dynamics are the cause and consequence of both chemical interactions and physical forces, and its analysis requires thinking at the interface of biochemistry and biomechanics [7].

In this chapter, we discuss a new approach to model cytoskeletal dynamics by focussing on the relatively simple context of filopodia. Filopodia are long finger-like membrane protrusions which act as tentacle-like sensors receiving signals as well as active devices that contact other cells to convey signals. The essential architectural feature of filopodia are parallel, cross-linked bundles of actin filaments which are pulled back into the cell through disassembly processes at their base, compensated for by extensive polymerization at the very tip. This leads to a continuous treadmill-like turnover of these actin filaments. By adjusting the proportion of the polar assembly versus disassembly processes, filopodia can undergo regulated length changes. However, the high rate of polymerization at the very tip of filopodia requires uninterrupted delivery of large amounts of new building blocks which, at first sight, seems counterintuitive in these long and slender protrusions [19, 22, 31, 33]. To explain this phenomenon, we propose the use of integrative models which consider catalyzed actin polymerization dynamics, diffusion as well as cytoplasmic flow dynamics.

2.2 The Object of Our Study

2.2.1 Actin Dynamics

Actin is the most abundant protein in eukaryotic cells [47]. Actin exists as globular actin monomers called G-actin and polar filaments called F-actin. Actin filaments are head-to-tail polymers of G-actin subunits. The minus, or pointed, end of actin filaments is relatively inert displaying slow growth in vitro. The opposite plus, or barbed, end grows much faster through exothermic polymerization both in vitro and in vivo [9].

In cells, actin filaments can be arranged into parallel bundles, (e.g., in finger-like membrane protrusions called filopodia), into anti-parallel bundles, (e.g., in stress fibers stretching across cells and acting as their contractile "muscle"), lattice-like networks of long filaments, (e.g., in lamellar cell protrusions called lamellipodia), carpet-like networks of short filaments, (e.g., in cortical actin underlying and structurally supporting the cell membrane), or as networks surrounding intracellular organelles [7]. The investigation of these different actin networks, in particular the flows and mechanical effects underlying plasma membrane protrusions, is an active field of experimentation and modeling, as is reviewed elsewhere [7, 15, 20, 38, 40].

As detailed in these reviews, there are three key features of actin networks (Fig. 2.1). Firstly, new filaments can be seeded de novo in a highly dynamic fashion through a process called nucleation, where several G-actins are catalytically oligomerized [10], from which state onwards they elongate efficiently through energy-favored polymerization. Secondly, actin filaments tend to be subjected to plus-end polymerization versus minus-end disassembly, i.e., they undergo treadmill process which translates into retrograde flow of whole actin networks if unengaged, or can become a source for force generation when linking to membrane or other cellular components [46]. Processes of polymerization and disassembly can be differentially and dynamically regulated through distinct classes of plus- and minus-end binding proteins, generating networks with different degrees of flow and constantly changing filament length. Thirdly, actin filaments can be bound one to

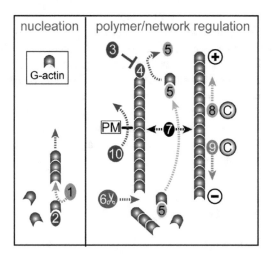

Fig. 2.1 Cartoon model of actin regulation. During nucleation, nucleators (*1*; e.g., Arp2/3, formins) cooperate with support factors (*2*; e.g., SCAR/WAVE, WASP, and APC) to generate linear actin oligomers which can then undergo exothermic polymerization. Plus-end polymerization is negatively regulated by capping proteins (*3*) that are competitively displaced by Ena/VASP (*4*) which, in turn, cooperates with profilin-G-Actin (*5*) in actin polymerization. F-actin disassembling factors (*6*; cofilin, myosin II) act at the minus-end. Actin cross-linking factors (*7*; e.g., fascin, actinin, and myosin II) assemble F-actin into networks or bundles and/or exert pulling forces. Different classes of myosin motors (*8, 9*) mediate cargo (C) transport towards the plus or minus-ends of the filaments. F-actin can be post-translationally modified (PM) by specialized enzymes

another by a range of different cross-linking proteins or myosin motor proteins; stabilizing filaments, weaving them into different classes of networks, using them as highways for transport processes or as scaffolds to pull and generate forces [7, 38].

Of particular importance for this chapter is the plus-end polymerization of actin. It occurs through intermolecular associations that depend on the special organization and molecular structure of the intervening monomers. Actin polymerization in cells is regulated by a number of proteins that bind G-actin and/or the actins at the plus-ends of the filaments. Amongst these, Ena/VASP, formins, profilin, and capping proteins are particularly important (Fig. 2.1). Capping proteins, such as CapZs or adducin, bind and stabilize the F-actin plus-end, i.e., suppress polymerization [1, 16].

Profilin binds and sequesters G-actin and is therefore, by default, an inhibitor of nucleation and enhancer of depolymerization [5]. However, this role changes dramatically if formins or Ena/VASP are present and bound to the elongating plus-ends. These proteins bind profilin-actin with high affinity and utilize it as G-actin source to catalyze and actively promote plus-end polymerization [3, 8]. In addition, they out-compete capping proteins, and Ena/Vasp can bundle the plus-ends of F-actin as well as position it at the membrane [3]. Therefore, apart from the physical properties of actin and its polymerization processes, the biochemical contributions of these actin regulators need to be considered in any models aiming to describe actin dynamics.

2.2.2 Filopodia

Filopodia are relatively simple cellular compartments providing a realistic context in which to start modeling the complexity of actin network regulation in biological contexts. Filopodia are long, finger-like membrane protrusions with numerous roles in signaling and cell navigation (Fig. 2.2a) [19, 22, 31, 33]. Unlike the enormous complexity of most cellular regions with dynamically interchanging actin networks, the comparatively simple organization of filopodia offers great advantages: (1) they contain only one form of prevailing actin network consisting of parallel F-actin bundles; (2) accordingly, the number of molecular players is limited; (3) filopodial dynamics are predominantly unidimensional, and (4) length changes of filopodial actin filament bundles directly translate into length changes of the entire filopodium, thus providing simple and efficient readouts for functional studies that can be carried out iteratively with modeling approaches (Fig. 2.2b–g) [44] .

Therefore, the regulation of filopodial dynamics appears relatively simple, essentially governed by the proteins regulating polymerization and disassembly processes. The key challenge in understanding filopodial dynamics is the high rate of polymerization at the very tip, which requires constant delivery of actin monomers through these slender and long structures. Therefore, how polymerization can be sustained within the highly challenging filopodial structure is a fascinating phenomenon which harbors key explanations for filopodial behaviors. Any models

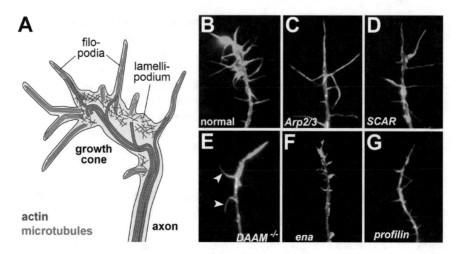

Fig. 2.2 Growth cone morphology. (**a**) A growth cone showing bundles of F-actin in filopodia and lattices in lamellipodia (taken from [55]). (**b–g**) Growth cones of fly neurons in culture lacking specific actin regulators which affect nucleation (**c–e**) leading to reduced filopodia number but not length, or polymerization (**f, g**) primarily affecting filopodial length (modified from [44])

of this process will have to consider that most G-actin is likely bound to profilin which will change its diffusion properties and enormously facilitate polymerization at the plus-end [3] where concentrations of G-actin can be expected to be very low. In support of this assumption, formins and Ena/VASP as active polymerizers and interactors of profilin are concentrated at filopodial tips, and deficiencies of profilin or Ena/VASP cause dramatic shortening of filopodia [44].

2.3 Suitable Biological Systems for Parallel Experimental Approaches

A key prerequisite for successful modeling is the existence of biological data and concepts/cartoon models. Since there is no single cellular system used in biological research which would alone provide sufficient insights and experimental data, the only feasible strategy to develop conceptual cartoon models for filopodial dynamics is to integrate mechanistic conclusions obtained from a wide range of very different cellular systems [19, 22, 31, 33]. However, this approach can be misleading since properties of filopodia may vary between animal species and cell types. Ideally, one would want a single and standardized experimental system so that data and mechanistic concepts can be reliably integrated. This would enormously facilitate the experimental validation of predictions made from mathematical modeling, especially if these predictions concern the interface between different parameters or elements.

As biological systems suitable for studying filopodia, growth cones are particularly well suited since they are rich in actin networks and reliably display filopodia as a prominent and functionally important structural feature (Fig. 2.2) [14, 23]. Growth cones are the motile tips of growing axons, the long and slender processes of nerve cells which form the cables that wire the nervous system. To lay these cables during development or regeneration, the growth cones at their tips navigate along specified paths, and their prominent filopodia act as sensors facilitating proper navigation [55]. Apart from the fact that growth cones display prominent filopodia with a clearly defined function, there is a good conceptual understanding of the fundamental roles that actin plays in growth cones, and their actin machinery is well investigated in this context [23, 41].

More recently, growth cones of the fruit fly *Drosophila* have been established as powerful models accessible to detailed studies of the cytoskeleton and systematic genetic dissections of its various regulators [49, 51, 55, 56]. For example, a systematic study used filopodial number and length as a simple readout to understand the systemic contributions and functional interfaces of seven different actin-binding regulators, which included formins, Ena/VASP, profilin, and capping proteins (Fig. 2.2b–g). This chapter demonstrated how *Drosophila* growth cones can be used to deduce cartoon models of filopodial dynamics [44, 50]. Given the relative speed and ease with which experiments can be carried out in *Drosophila* neurons, this system is ideal for partnering up with the mathematical/computational modeling of filopodial dynamics [49].

2.4 Modeling Filopodia

2.4.1 Previous Models

Since the end of the 1950s, researchers have been developing models to quantitatively validate experimental results from their laboratory experiments. Very early on, Oosawa and colleagues found that actin polymerization processes were dependent of G-actin concentration [39]. In another early seminal study, Wegner mathematically described actin treadmilling [58]. A further good example is the ordinary differential equation model by Bindschadler and colleagues which explicitly accounts for nucleotide-dependent actin polymerization and depolymerization [4].

At about the same time, taking advantage of technological advance and detailed experimentally measured data, Vavylonis and colleagues built a quantitative discrete model of actin polymerization [57]. This model could be validated by new in vitro experiments using reflection fluorescence microscopy [21]. Mechanisms of actin polymerization have been studied including the interactions between actins and actin-binding proteins (ABPs) and the rate constants for each step leading to quantitative models of polymerization [47].

Another important model that elucidated thinking about how actin polymerization can dynamically control cell shape and motility was the Brownian ratchet model by Peskin and colleagues [43] which aims to explain how chemical reactions generate protrusive forces by rectifying Brownian motion. The ratchet mechanism is the intercalation of monomers between the barrier, (e.g., membrane) and the polymer tip. A particle diffuses in one dimension ahead of a growing bundle of filaments, executing a continuous random walk in a constant force field. As a result, it analyzes polymerization velocity as a function of the load force.

Actin dynamics in motile cells have also been the subject of many modeling approaches. For example, the model by Mogilner and Keshet showed how membrane speed depends on actin monomer concentration and cycle (ATP hydrolysis, F-actin treadmilling), reactions with ABPs and barbed end polymerization at the leading edge of a cell [34]. Dawes and colleagues investigated the spatial distribution of actin filaments and their barbed ends, and the interplay between filament branching, growth, and decay at the leading edge [12]. Others investigated the mechanisms of pushing and pulling by actin and microtubules [35], the relations between protrusions, monomer concentration, and stiffness of the filament bundles [36], or anterograde flow of cytoplasm to provide sufficient material for the extension of protrusions, such as lamellipodia [27].

As these examples illustrate, the modeling of actin in biological contexts has to include different levels of resolution, such as structural, physical, and chemical properties of actin filaments and protein complexes that govern F-actin dynamics [7].

With regard to modeling filopodia, several previous attempts should be mentioned. For example, Lan and Papoian used stochastic simulations of filopodial dynamics, discretizing space into compartments, and simulating protein diffusion along the filopodium as a random walk [28]. Erban and colleagues modeled the problem of G-actin delivery to the filopodial tip based on diffusion, 3D stochastic models, and comparing compartmental and molecular models as strategies to simulate actin dynamics [17]. An impressively comprehensive model was published by Mogilner and Rubinstein who included aspects such as physical properties, cross-linkage and numbers of filaments, membrane resistance as well as the G-actin diffusion coefficient [36]. Another recent model by Zhuravlev and colleagues proposed that a key limiting factor of filopodial length is diffusional transport of G-actin monomers to the polymerizing barbed ends, and they investigated potential roles of active motor-driven transport of G-actin [61]. They concluded that "a naive design of molecular-motor-based active transport would almost always be inefficient—an intricately organized kinetic scheme, with finely tuned rate constants, is required to achieve high-flux transport."

2.4.2 Can Diffusion Alone Explain G-Actin Delivery to the Filopodial Tip?

We first explored whether, and under which conditions, diffusion might be sufficient to supply barbed end actin polymerization processes in filopodia considering the work by others who investigated actin diffusion as a key mechanism in filopodia. For reasons of simplicity, our calculations do not consider filopodial elongation, but analyze conditions where retrograde flow and polymerization are in balance, thus maintaining filopodial shape, length, retrograde flow in a steady state. For our calculations we used the parameter values indicated in bold text in Table 2.1.

In short filopodia of $1–2\,\mu m$, the concentration of free actin at the filament base is still sufficient to supply enough polymerization for filopodial extension to occur [36]. This is in agreement with (2.1) for estimating the approximate time required for a particle to diffuse over a given distance x, in an environment where its diffusion coefficient is D

$$t \approx \frac{x^2}{q_i D} \tag{2.1}$$

Table 2.1 Parameters and data regarding filopodia and actin

Notation	Meaning	Value	References
L	Filopodial length	$24–55\,\mu m$	[2]
		$0.03–0.15\,\mu m$	[35]
		$1–10\,\mu m$	[43]
		$10–20\,\mu m$	[52]
		$3–10\,\mu m$	[56]
		Here: **$1–30\,\mu m$**	
N	Number of filaments in the filopodial bundle	$10–30$	References in [36]
$\mathbf{C_0}$	G-actin concentration at filopodial base	$10\,\mu M$	[36]
k_{on}	Polymerization rate	$10\,\mu M^{-1}s^{-1}$	[36]
		$11.6\,\mu M^{-1}s^{-1}$	[60]
N_0	Number of filaments to support protrusion	13	[36]
η	Unit conversion factor	20	[36]
δ	Half-size of actin monomer	$2.7\,nm$	References in [43]
D	G-actin diffusion coefficient	$4\,\mu m^2/s$	[35]
		$5\,\mu m^2/s$	[32]
V_{ret}	Retrograde flow rate	**$70\,nm/s$**	[36]
		$30–80\,nm/s$	[56]

Values selected for this chapter are shown in bold

In (2.1), q_i is 2, 4, or 6 depending on the number of dimensions ($i = 1, 2$ or 3) [26]. Considering a linear displacement ($q_i = 2$) and a diffusion coefficient of $5\,\mu m^2/s$, a G-actin travels $3.16\,\mu m/s$.

For filaments longer than a few micrometers, we can analyze concentration of free actin according to expression (2.2) mentioned in [36]:

$$C(x) = C_0 - \frac{C_0 x}{L(t) + (D\eta e^{N_0/N})/(k_{on}N)} \qquad (2.2)$$

In (2.2), $L(t)$ indicates filopodial length as a function of time, (i.e., considering filopodial elongation or shrinkage). Our work does not consider elongation, so we assume $L(t) = L$ constant. Furthermore, we chose the filopodial length interval to cover for a wide range of natural filopodia (Table 2.1), although values between 10 and $20\,\mu m$ are more frequently observed [7]. Function $C(x)$ is the G-actin concentration at a given distance between the filopodial base ($x = 0$) and the filopodial tip ($x = L$). Therefore, the free actin concentration at the tip of the filopodium is given by the function expressed in

$$C = C(L, N, k_{on}) = C_0 - \frac{C_0 L}{L + (D\eta e^{N_0/N})/(k_{on}N)} \qquad (2.3)$$

This value allows for calculating how many actins reach the polymerization point N_p in scenarios generated by varying L, N, and k_{on}. The chosen values are in the ranges reported in Table 2.1.

The key question is whether diffusion is sufficient to sustain polymerization in order to balance the actin-bundle retrograde flow ($V_{ret} = 70$ nm/s in Table 2.1). To achieve this velocity, a polymerization frequency of 30 actins/s per filament in the bundle is required when considering that every new actin elongates a filament by ≈ 2.7 nm (Table 2.1). Therefore, in a bundle with N filaments, $N_{ret} = 30\,N$ actins should be polymerized per second. This will be compared with the number of actins obtained from (2.3), to check if diffusion is enough to sustain F-actin retrograde flow.

At a G-actin concentration of $10\,\mu M$, the polymerization rate per actin filament was reported to be $0.3\,\mu m/s$, which means that 110 actins are polymerized in a single filament per second [45]. For a bundle with N filaments, the number of polymerizing actins N_p would be $110\,N/s$. By integrating this with expression (2.3), we have

$$NP = 110NC \qquad (2.4)$$

as the number of actin polymerized at a given concentration $C = C(L, N, k_{on})$. The calculations were performed by the following routine:

```
for L in range(min = 0.5, max = 30, step = 0.5):
    for N in range(min = 10, max = 30, step = 1):
        for k_on in range(min = 10, max = 12, step = 1):
            Calculate C = C(L, N, k_on), following (2.3)
```

$$N_{\text{ret}} = 30N$$
$$N_p = 11NC$$
If $N_p < N_{\text{ret}}$
    Add one in the number of cases where $N_p < N_{\text{ret}}$
else
    Add one in the number of cases where $N_p \geq N_{\text{ret}}$ print 'Number of cases
where $N_p < N_{\text{ret}}$ (in %)'

Here $N_p = 11NC$, because the number of polymerized actins is directly proportional to the concentration of monomers. At a concentration of $10\,\mu$, 110 actins are polymerized per second on each filament, providing 11 actins/s at $1\,\mu$M. The expression (3) yields $C \to C_0$ when $L \to 0$, ensuring that $N_p = 110N$ for a concentration close to that at the base of filopodia.

Using this algorithm in combination with the values given in Table 2.1, we obtain that of all cases the polymerization does not supply retrograde flow in 90.32 %. In comparison, when setting the filopodial length interval to $L \in [0.5, 2]\,\mu$m, diffusion seems ineffective in just 15 % of cases. This result suggests that diffusion alone is not sufficient in many scenarios, especially when reaching filopodial lengths beyond $1.5\,\mu$M, and when N is greater than 21.

Other reports agree with our view. For example, Monte Carlo simulations were used to investigate G-actin translocation during protrusion of the leading edge leading to the conclusion that diffusion alone was insufficient [62]. The use of compartmental and molecular stochastic models to study actin motion by diffusion concluded that filopodia would reach a steady state length of as little as $\approx 1\,\mu$m because the transport flux of G-actin monomers continuously diminishes as the tube becomes longer [17]. Also work on filopodia-like acrosomal processes of sperm found that the kinetics of diffusion-limited actin polymerization were not sufficiently rapid to account for the observed acrosomal elongation dynamics [42].

2.4.3 An Alternative Approach

The above calculations show that diffusion alone is, in general, not enough to supply the amount of G-actin needed to sustain the observed rates of polymerization and backflow. As one compensatory mechanism, Zhuravlev and colleagues proposed active motor-driven transport of G-actin [61], but there is hardly any experimental proof for this hypothesis. Therefore, we propose in the sequel a modeling approach which integrates a number of very different physical and chemical processes within filopodia (Fig. 2.3a): diffusion of actin building blocks, support-factor aided polymerization at the filament tips, as well as cytoplasmic flow driven by the forces caused by the displacement of actin filaments towards the filopodial base.

The idea of protein-mediated facilitation of polymerization was already explained in Sect. 2.2.1. Here, we briefly explain the concept of cytoplasmic flow. Thus, F-actin filaments at the core of the filopodium are constantly flowing backwards from the tip towards the cell through disassembly processes occurring

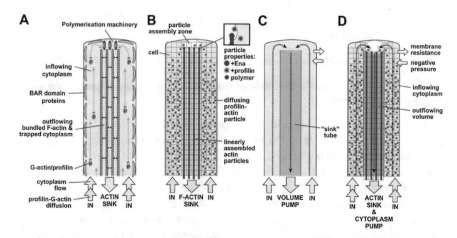

Fig. 2.3 Cartoon models of filopodial dynamics. (**a**) Biological model: in cells, actin bundles are pulled out of filopodia from their minus-ends (ACTIN SINK) which is compensated through Ena/VASP-mediated (E) plus-end polymerization requiring supply (*yellow arrows*) with profilin (P)-bound G-actin (same symbols as explained in Fig. 2.1). Filopodial membranes are stabilized by BAR domain proteins. (**b–d**) Mathematical/computational models: (**b**) A model of diffusion and polymerization: the filopodium is subdivided into a coordinate system within which actin particles (*magenta dots*) undergo Brownian movement, diffusion, and assembly processes; to reflect contributions by regulatory proteins, actin particles are equipped with profilin- or Ena/VASP-bound properties (*orange or brown circles*; see close-up top right). (**c**) Flow model: filopodia are represented by a tube-in-tube constellation where volume (actin filaments with potentially trapped cytoplasm) flows out of filopodia, driven by a pump at the base of the inner tube, causing compensatory inflow of cytoplasm. (**d**) Combinatorial model: the coordinate system becomes dynamic reflecting the circulating cytoplasm flow from (**c**); actin particles with different properties (as in **b**) can diffuse and polymerize within this dynamic context, thus combining cytoplasmic flow, diffusion, and regulator-mediated polymerization

within the cell (Fig. 2.3a). This causes loss of volume in the filopodial tip which comprises the actin filaments, their hydration, other attached proteins, and perhaps even cytoplasm trapped in between the tightly packed actin filaments. If this volume loss is combined with sufficiently resistant membrane structure to prevent membrane collapse (open arrow in Fig. 2.3c), this would generate a negative pressure that needs to be compensated for by incoming cytoplasm to prevent cavitation.

Indeed filopodia seem to display such membrane rigidity provided by a class of proteins containing BAR (Bin-Amphiphysin-Rvs) domains (Fig. 2.3a) [30, 53, 54]. Given therefore a relatively stable shape of filopodia, the "outflow" of volume can be expected to drive a compensatory influx in the space between actin bundle and cell membrane towards the tip of the filopodium (Fig. 2.3c). The inflowing volume should be a mixture of cytosol (colloidal water) and G-actin molecules where the latter are still free to diffuse (Fig. 2.3d). At the tip of the filament the mixture-flow bends in the direction of the polymerization points, guided by the molecular organization around it, where the diffused G-actins are re-arranged through polymerization into the backflowing filament bundle. The bending flow might even produce outward forces in the membrane that could further help sustain its form.

2.5 A Model for Filopodial Dynamics

2.5.1 Principal Thoughts About Strategic Choices

Three processes are central in the above description: diffusion, guided polymerization, and cytoplasmic flow. With polymerization being a chemically reactive process, we have a diffusion–reaction system with advection. In principle, there are several standard models from several mathematical disciplines which can describe such processes.

However, the mathematical modeling approaches to implement advection dynamics are very different from, and complementary to, those required by reaction–diffusion. Despite its hydro-dynamical elegance, the alternative approach above, that suggests a reorganization of particles within a flowing cytosol, cannot arise under the commonly used diffusion perspective because cytosol displacements are disregarded and only diffusive movements are represented. A new starting point is needed to suggest the models and observations needed to test it.

Furthermore, setting the standard equations for reaction, diffusion, and advection *per se* down as a model for filopodial dynamics is challenging, because they were developed based on very basic principles of homogeneity, uniformity, and isotropy, and the substances that react or drift are supposed to be present at any point of the domain. However, filopodia are anything else than isotropic or uniform. Different processes are predominant in very distinct regions of the filopodia and, thus, different equations should be used in distinct subregions of the model's domain. For example, the standard equations allow, indeed enforce, actin reactions to take place anywhere in the filopodium and consider actin molecules as uniformly charged spheres, hampering the representation of what we know about protein-aided polymerization reactions which occur only at the plus-end of each filament. Furthermore, filopodia are bounded by (1) the moving cell membrane and (2) a rather unspecified surface at the bottom of the protrusion, and (3) further sub-domains are formed by the dynamic flow of distinct molecule classes: the backflowing actin bundles versus the (laminar) advection outside them. Therefore, the geometrical boundaries of each sub-domain cannot be determined in advance, as they result from molecular dynamics. Taking into consideration that we need to discretize the equations to perform simulations and numerical experiments, and that we are dealing at the very validity limit of continuous flow models [29], our chance of getting close to helpful and interesting solutions in this way is not very likely.

Based on these considerations, we prefer therefore computational models based on both mathematical and biological knowledge and tailored to what is known to happen in each sub-domain of a filopodium. We consider here the combined use of two computational methods. First, the finite volume discretization with moving elements method, based on standard models of laminar flow and discretization methods which can track changes in the sub-domain geometry [13]. Second, the particle-in-cell method (PIC) relating to reaction–diffusion equations. This method can be used to model stochastic processes and simultaneously represent the rigid motion of the F-actin filament bundle, simply by considering transitions to be deterministic and remaining in a direction given by the central axis of the cylindrical portion of the filopodium.

2.5.2 Modeling F-Actin Polymerization and Diffusion

The PIC method we chose is a numerical technique that deals with problems where a fluid can be seen as particles scattered in a field described by a mesh, cf. [18, 48]. Every grid cell has a sufficiently large quantity of particles interacting with each other, carrying information about its own position and physico-chemical properties. PIC exhibits a widespread usage in computer simulation of plasma physics, gases, or other fluids composed of particles with high kinetic energy. It seems an efficient strategy to simulate the two-phase mixture here described.

Filopodia contain a collection of interacting particles and the cytosol. By particles we mean G-actin, F-actin, and ABPs, i.e., the key factors that interact and regulate the entire system. They are placed over a background field formed by the cytosol fluid or cytoplasm. Our model domain is an environment that involves a multi-scale fluid observation, in order to consider relevant, microscopic, and physical phenomena that bear influence on macroscopic properties of fluids.

In our chosen PIC model, each particle's position is updated in response to system stimuli, such as fluid dragging, collision with other particles, or chemical reactions. For example, polymerization processes are triggered by the proximity of G-actins with the tip of F-actin bundles to which they bind. In this regime, distance can be used as a priority-parameter to decide which molecule polymerizes (Fig. 2.4).

In addition, the position-tracking of particles permits to measure deformations through observation of displacements of F-actin position (Fig. 2.5). This information will be relevant in future analyses, for example, when incorporating forces, F-actin buckling, and membrane resistance.

In our preliminary simulations, particles move in Brownian motion through a 2D domain, and we observe G-actin concentration depletion in the area around a polymerization point (red dot in Fig. 2.6). This is an expected behavior since G-actins are consumed by polymerization and Brownian motion of other particles is not efficient enough to fill the vacant spaces.

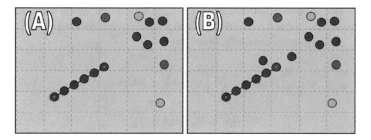

Fig. 2.4 Schemes to represent possibility of particle interactions in our PIC simulations. (**a**) Several particles can be moving within same tiles (**b**) If a G-actin is in the same tile of polymerization point, it can be polymerized

Fig. 2.5 Deformation of the filament measured through changes in particle positions

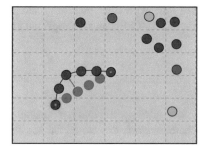

Fig. 2.6 Particles (G-actins, *blue*) moving in Brownian motion, F-actins in *green* with the polymerization point (*red dot*) from where these filaments elongate. Note the decreasing G-actin concentration around the polymerization point

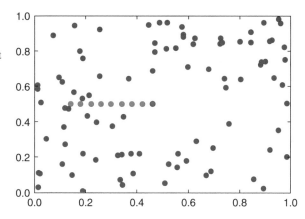

2.5.3 *Modeling Cytosolic Flow as a Means to Drive Passive Transport of G-Actin*

In order to investigate how G-actins are delivered to the polymerization points, we must consider physical properties of cytosol and fluctuations in the concentration of G-actins and other molecules throughout the domain. Therewith, we analyze cellular compensatory mechanisms for replacing the molecules consumed by polymerization, subject to the principle of mass conservation. Cytosol has a colloidal-water constitution which, in first approximation, leads to considering it as an incompressible and viscous fluid and modeling its motion by Navier–Stokes equations for incompressible fluid flow [29],

$$\rho \left(\frac{\partial \mathbf{U}}{\partial t} + (\mathbf{U} \cdot \nabla)\mathbf{U} \right) = -\nabla p + \mu \mathbf{U} + \mu \nabla^2 \mathbf{U} + F \tag{2.5}$$

$$\frac{\partial \rho}{\partial t} + \nabla \cdot (\rho \mathbf{U}) = F_m \tag{2.6}$$

Here $\mathbf{U}$ stands for the velocity vector, ρ for the fluid density, μ for the dynamic viscosity, p for the hydrostatic pressure, F for the source of movement, and finally F_m for the source of mass. Considering filopodia as a system where fluid velocities are low, with high viscosity, and very small length-scales, Navier–Stokes above Eqs. (2.5)–(2.6) can be simplified and written as:

$$\nabla p - \mu \nabla^2 \mathbf{U} = F \tag{2.7}$$

$$\nabla \mathbf{U} = F_m \tag{2.8}$$

taking into account that ρ = constant, due to fluid incompressibility. Flows with these characteristics are called Stokes flows and Eqs. (2.7)–(2.8) are known as Stokes equations.

At this point, the fluid can be analyzed through a macroscopic approach, i.e., in a laminar flow with a certain concentration of G-actins in suspension, since the latter are supposed to be passively transported by the cytosol and do not affect its flow. This scenario is based on assumptions described in previous sections: the volume removed from the filopodia tip induces a hydrostatic pressure, thus leading to the cytosol displacement.

The conditions and processes we seek to investigate with the tools under development include the rules underlying G-actin motion, hydrostatic pressure, and membrane resistance. Aiming to enforce mass conservation inside the filopodia, which is a critical aspect of our approach, we have decided to discretize the model flow equations with the finite volume method. Notably, PIC can be coupled with finite volume laminar flow giving an integrated two-phase representation of an entire system as a dynamic fluid, to study the reorganization of the actin-phase into filaments at molecular scales. Within a neighborhood of the polymerization points, the two methods merge by superposition: the PIC method determines the movement of G- and F-actins relative to the volume elements, and the finite volume algorithm drives the displacement and deformation of these volume elements. Volume elements trapped within the F-actin filament bundle move then with the filaments.

The domain of our model mimics filopodia interior and will initially be two-dimensional for implementation tests. The scheme in Fig. 2.7 represents a filopodium and a bundle of actin filaments inside, where red arrow means retrograde flow. The cytoplasmic fluid is pumped from lamellipodia in $x = 0$,

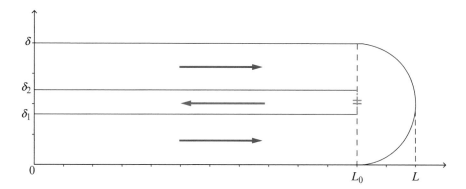

Fig. 2.7 Filopodia fluid motion scheme

for $y \in [0, \delta_1) \cup (\delta_2, \delta]$. Filopodium is also bounded by the membrane at $x = L$. Actin polymerization occurs at $x = L_0$ where $y \in [\delta_1, \delta_2]$. This region is the inset of a tube that represents F-actin filament bundle and drives actins out of filopodia, acting as a sink.

2.5.4 Model Validation

Model validation is a long-term project which relies on the confidence we gain from solving questions and explaining biological measurements with our model. Nevertheless, there are series of immediate experiments that can be undertaken to check both the model's conceptual soundness and its implementation.

We have seen in Sect. 2.5.2 that the PIC model represents well the diffusion of G-actins, the formation of filaments, and their almost rigid-like motion along a line. Computational models for laminar flows based on finite volumes are well known, as well as their validation. The steps below address the coupling of both algorithms and the representation of characteristics relevant for biological enquiries and the special filopodia geometry.

First, we need to verify basic aspects: mass conservation, soundness of the flow bending at the tip, coupling of advective currents with the polymerization points, and the return of G-actins by the cytosolic inward flow. Furthermore, we need to verify if the superimposition of both models retains properties relevant for the biological problem: does it (1) provide enough actins at the polymerization points? (2) reduce the flow when polymerization stops? (3) represent elongation and retraction of the filopodia? and so on.

Moreover, the model can be used to perform tests that provide data supporting the approach proposed, like those suggested in Fig. 2.8. The results of numerical (virtual) experiments of this type provide data in favor or against the proposed approach.

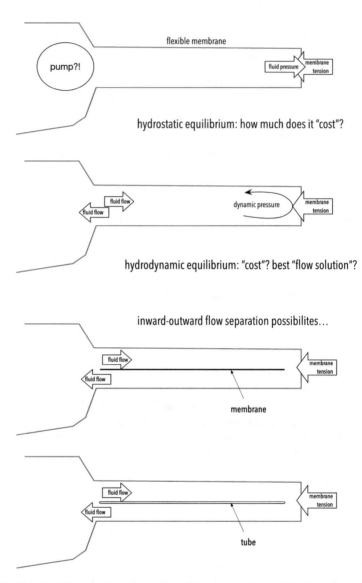

Fig. 2.8 Biological Quests—what is the most efficient way of sustaining the cell membrane in protrusions, and providing enough actins to keep the polymerization rate?

2.6 Conclusions

This chapter presents a novel approach to the description of the flow of matter inside filopodia, adding advection of the cytosol to the diffusion of molecules suspended or dissolved within it. Proposing this advection is justified by the biological knowledge that volume (actin, attached proteins, and cytosol) is withdrawn from the filopodia, providing a pump mechanism for compensatory inflow. This research is in its infancy and a lot of open questions lay ahead, besides those related to model validation. These open questions include: (1) relations between the number of filaments in the bundle and several aspects, like flow velocity, G-actin concentration, elongation–retraction of the protrusion, etc; (2) the mechanical effects or the flow over the membrane throughout the filopodia; and (3) how the outward movement of the filament bundle affects the laminar inward flow. We are working towards joining the PIC and the finite volume algorithms and hope to be addressing these intriguing questions in the near future and couple them up with imaging experiments in *Drosophila* growth cones. However, our ultimate goal is to understand how this flow apparatus can be regulated and controlled to yield all behavior we observe in filopodia.

Acknowledgements This chapter has been conducted as a partial result of an agreement between Faculty of Life Sciences (FLS), University of Manchester, and Brazilian National Laboratory of Scientific Computing (LNCC), in collaboration with researchers from Rio de Janeiro State University—UERJ and Federal Institute of Rio de Janeiro—IFRJ. Special thanks go to UK Biotechnology and Biological Sciences Research Council (BBSRC) for sponsoring project BB/L026724/1, to FAPERJ for project APQ-5 E-26/110.931/2014, UERJ-SR2 Visiting Researcher 2014–2016 grant, and a CNPq/DTI grant.

The authors express their gratitude to the editors for suggestions to reformulate and deeply improve this chapter.

References

1. Baines, A.J.: The spectrin-ankyrin-4.1-adducin membrane skeleton: adapting eukaryotic cells to the demands of animal life. Protoplasma **244** (2010). http://www.ncbi.nlm.nih.gov/pubmed/20668894
2. Bathe, M., Heussinger, C., Claessens, M.M., Bausch, A.R., Frey, E.: Cytoskeletal bundle mechanics. J. Biophys. **94**, 2955–2964 (2008)
3. Bear, J.E., Gertler, F.B.: Ena/vasp: towards resolving a pointed controversy at the barbed end. J. Cell Sci. **122** (2009). http://www.ncbi.nlm.nih.gov/entrez/query.fcgi?cmd=Retrieve&db=PubMed&dopt=Citation&list_uids=19494122
4. Bindschadler, M., Osborn., E.A., McGrath, J.L.: A mechanistic model of the actin cycle. Biophys J. 2004 May; **86**(5), 2720–2739. doi: 10.1016/S0006-3495(04)74326-X
5. Birbach, A.: Profilin, a multi-modal regulator of neuronal plasticity. Bioessays **30** (2008). http://www.ncbi.nlm.nih.gov/entrez/query.fcgi?cmd=Retrieve&db=PubMed&dopt=Citation&list_uids=18798527
6. Blagoev, K.B., Shukla, K., Levine, H.: We need theoretical physics approaches to study living systems. Phys. Biol. **10**, 040,201 (2pp) (2013). doi:10.1088/1478-3975/10/4/040201. http://www.ncbi.nlm.nih.gov/pubmed/23883648
7. Blanchoin, L., Boujemaa-Paterski, R., Plastino, C.S.J.: Actin dynamics, architecture, and mechanics in cell motility. Physiol. Rev. **94** (2014). http://www.ncbi.nlm.nih.gov/pubmed/24382887
8. Breitsprecher, D., Goode, B.L.: Formins at a glance. J. Cell Sci. **126**, 1–7 (2013). http://jcs.biologists.org/content/126/1/1.full.pdf+html
9. Carlsson, A.E.: Actin dynamics: from nanoscale to microscale. Ann. Rev. Biophys. **39**, 91–110 (2010). doi:10.1146/annurev.biophys.093008.131207

10. Chesarone, M.A., Goode, B.L.: Actin nucleation and elongation factors: mechanisms and interplay. Curr. Opin. Cell Biol. **21**, 28–37 (2009)
11. Cohen, J.E.: Mathematics is biology's next microscope, only better; biology is mathematics' next physics, only better. PLoS Biol. **2**, e439 (2004). http://www.ncbi.nlm.nih.gov/pubmed/15597117
12. Dawes, A.T., Ermentrout, G.B., Cytrynbaum, E.N., Edelstein-Keshet, L.: Actin filament branching and protrusion velocity in a simple 1d model of a motile cell. J. Theor. Biol. **242**, 265–279 (2006)
13. Demirdzic, I., Peric, M.: Finite volume method for prediction of fluid flow in arbitrarily shaped domains with moving boundaries. Int. J. Numer. Methods Fluids **10**, 771–790 (1990)
14. Dent, E.W., Gupton, S.L., Gertler, F.B.: The growth cone cytoskeleton in axon outgrowth and guidance. Cold Spring Harb. Perspect. Biol. **3**, a001,800 (2011). http://www.ncbi.nlm.nih.gov/entrez/query.fcgi?cmd=Retrieve&db=PubMed&dopt=Citation&list_uids=21106647
15. Ditlev, J.A., Mayer, B.J., Loew, L.M.: There is more than one way to model an elephant. experiment-driven modeling of the actin cytoskeleton. Biophys. J. **104**, 520–532 (2013)
16. Edwards, M., Zwolak, A., Schafer, D.A., Sept, D., Dominguez, R., Cooper, J.A.: Capping protein regulators fine-tune actin assembly dynamics. Nat. Rev. Mol. Cell Biol. **15**, 677–689 (2014). http://dx.doi.org/10.1038/nrm3869
17. Erban, R., Flegg, M.B., Papoian, G.A.: Multiscale stochastic reaction-diffusion modeling: application to actin dynamics in filopodia. Bull. Math. Biol. **76**, 799–818 (2014). http://www.ncbi.nlm.nih.gov/pubmed/23640574
18. Evans, M., Harlow, F.H.: The particle-in-cell method for hydrodynamic calculations. Technical Report, LA-2139, 73p. Los Alamos Scientific Laboratory, New Mexico (1957)
19. Faix, J., Breitsprecher, D., Stradal, T.E.B., Rottner, K.: Filopodia: complex models for simple rods. Int. J. Biochem. Cell Biol. **41**(8–9), 1656–1664 (2009). doi:10.1016/j.biocel.2009.02.012
20. Fletcher, D.A., Mullins, R.D.: Cell mechanics and the cytoskeleton. Nature **463**, 485–492 (2010). http://www.ncbi.nlm.nih.gov/pubmed/20110992
21. Fujiwara, I., Vavylonis, D., Pollard, T.D.: Polymerization kinetics of adp- and adp-pi-actin determined by fluorescence microscopy. Proc. Natl. Acad. Sci. USA **104**, 8827–8832 (2007). http://www.ncbi.nlm.nih.gov/pubmed/17517656
22. Gallo, G.: Mechanisms underlying the initiation and dynamics of neuronal filopodia: from neurite formation to synaptogenesis. Int. Rev. Cell. Mol. Biol. **301**, 95–156 (2013). http://www.ncbi.nlm.nih.gov/pubmed/23317818
23. Gomez, T.M., Letourneau, P.C.: Actin dynamics in growth cone motility and navigation. J. Neurochem. **129**(2), 221–234 (2013). http://www.ncbi.nlm.nih.gov/pubmed/24164353
24. Gunawardena, J.: Beware the tail that wags the dog: informal and formal models in biology. Mol. Biol. Cell. **25**, 3441–4 (2014). http://www.ncbi.nlm.nih.gov/pubmed/25368417
25. Gunawardena, J.: Models in biology: 'accurate descriptions of our pathetic thinking'. BMC Biol. **12**, 29 (2014). doi:10.1186/1741-7007-12-29. http://www.biomedcentral.com/1741-7007/12/29
26. Islam, M.A.: Einstein-Smoluchowski diffusion equation: a discussion. Phys. Scripta **70**, 120–125 (2004)
27. Kucik, D.F., Elson, E.L., Sheetz, M.P.: Forward transport of glycoproteins on leading lamellipodia in locomoting cells. Nature **340**, 315–317 (1989)
28. Lan, Y., Papoian, G.: The stochastic dynamics of filopodial growth. Biophys. J. **94**, 3839–3852 (2008)
29. Liu, C., Zhigang, L.: On the validity of the navier-stokes equations for nanoscale liquid flows: the role of channel size. AIP Adv. **1**, 032,108 (2011). doi:10.1063/1.3621858
30. Masuda, M., Mochizuki, N.: Structural characteristics of bar domain superfamily to sculpt the membrane. Semin. Cell Dev. Biol. **21**, 391–398 (2010). http://www.ncbi.nlm.nih.gov/pubmed/20083215
31. Mattila, P.K., Lappalainen, P.: Filopodia: molecular architecture and cellular functions. Nat. Rev. Mol. Cell. Biol. **9**, 446–54 (2008). http://www.ncbi.nlm.nih.gov/entrez/query.fcgi?cmd=Retrieve&db=PubMed&dopt=Citation&list_uids=18464790

32. McGrath, J.L., Tardy, Y., Dewey, C.F., Meister, J.J., Hartwig, J.H.: Simultaneous measurements of actin filament turnover, filament fraction, and monomer diffusion in endothelial cells. Biophys. J. **75**, 2070–2078 (1998)
33. Mellor, H.: The role of formins in filopodia formation. Biochim. Biophys. Acta **1803**(2), 191–200 (2009). doi:10.1016/j.bbamcr.2008.12.018. http://www.ncbi.nlm.nih.gov/entrez/query.fcgi?cmd=Retrieve&db=PubMed&dopt=Citation&list_uids=19171166
34. Mogilner, A., Edelstein-Keshet, L.: Regulation of actin dynamics in rapidly moving cells: a quantitative analysis. Biophys. J. **83**, 1237–1258 (2002)
35. Mogilner, A., Oster, G.: Cell motility driven by actin polymerization. Biophys. J. **71**, 3030–3045 (1996). http://www.ncbi.nlm.nih.gov/pubmed/8968574
36. Mogilner, A., Rubinstein, B.: The physics of filopodial protrusion. Biophys. J. **89**, 782–95 (2005). http://www.ncbi.nlm.nih.gov/pubmed/15879474
37. Mogilner, A., Allard, J., Wollman, R.: Cell polarity: quantitative modeling as a tool in cell biology. Science **336**, 175–179 (2012). http://www.ncbi.nlm.nih.gov/pubmed/22499937
38. Mullins, R.D., Hansen, S.D.: In vitro studies of actin filament and network dynamics. Curr. Opin. Cell Biol. **25**, 6–13 (2013). http://www.sciencedirect.com/science/article/pii/S0955067412001901
39. Oosawa, F., Asakura, S., Ooi, T.: G-f transformation of actin as a fibrous condensation. J. Polym. Sci. **37**, 323–336 (1959). http://dx.doi.org/10.1002/pol.1959.1203713202
40. Othmer, H.: Actin cytoskeleton, multi-scale modeling. In: Engquist, B. (ed.) Encyclopedia of Applied and Computational Mathematics, pp. 17–23. Springer, Berlin/Heidelberg (2015). doi:10.1007/978-3-540-70529-1_60
41. Pak, C.W., Flynn, K.C., Bamburg, J.R.: Actin-binding proteins take the reins in growth cones. Nat. Rev. Neurosci. **9**, 136–47 (2008). http://www.ncbi.nlm.nih.gov/entrez/query.fcgi?cmd=Retrieve&db=PubMed&dopt=Citation&list_uids=18209731
42. Perelson, A.S., Coutsias, E.A.: A moving boundary model of acrosomal elongation. J. Math. Biol. **23**, 361–379 (1986)
43. Peskin, C.S., Odell, G.M., Oster, G.F.: Cellular motions and thermal fluctuations: the Brownian ratchet. Biophys. J. **65**, 316–324 (1993)
44. Pimentel, C.G., Gombos, R., Mihály, J., Sánchez-Soriano, N., Prokop, A.: Dissecting regulatory networks of filopodia formation in a Drosophila growth cone model. PLoS ONE **6**, e18,340 (2011). doi:10.1371/journal.pone.0018340. http://www.plosone.org/article/info%3Adoi%2F10.1371%2Fjournal.pone.0018340
45. Pollard, T.D.: Rate constants for the reactions of atp-and adp-actin with the ends of actin filaments. J. Cell Biol. **103**, 2747–2754 (1986)
46. Pollard, T.D.: Regulation of actin filament assembly by arp2/3 complex and formins. Ann. Rev. Biophys. Biomol. Struct. **36**, 451–477 (2007)
47. Pollard, T.D., Cooper, J.A.: Actin and actin-binding proteins. a critical evaluation of mechanisms and functions. Ann. Rev. Biochem. **55**, 987–1035 (1986)
48. Potter, D.: Computational Physics. Wiley, New York (1973)
49. Prokop, A., Beaven, R., Qu, Y., Sánchez-Soriano, N.: Using fly genetics to dissect the cytoskeletal machinery of neurons during axonal growth and maintenance. J. Cell Sci. **126**, 2331–41 (2013). doi:10.1242/jcs.126912
50. Prokop, A., Sánchez-Soriano, N., Calves Pimentel, C.G., Molnár, I., Kalmr, T., Mihály, J.: Daam family members leading a novel path into formin research. Commun. Integr. Biol. **4**, 538–542 (2011). http://www.landesbioscience.com/journals/cib/article/16511/
51. Prokop, A., Küppers-Munther, B., Sánchez-Soriano, N.: Using primary neuron cultures of *Drosophila* to analyse neuronal circuit formation and function. In: Hassan, B.A. (ed.) The making and un-making of neuronal circuits in *Drosophila*, vol. 69, pp. 225–247. Humana Press, New York (2012). doi:10.1007/978-1-61779-830-6_10. http://www.springerlink.com/content/t07618161235u475/#section=1102403&page=1
52. Pronk, S., Geissler, P.L., Fletcher, D.A.: Limits of filopodium stability. Phys. Rev. Let. **100**, 258,102 (2008)

53. Rao, Y., Haucke, V.: Membrane shaping by the Bin/Amphiphysin/Rvs (BAR) domain protein superfamily. Cell Mol. Life Sci. **68**, 3983–3993 (2011). http://www.ncbi.nlm.nih.gov/pubmed/21769645

54. Safari, F., Suetsugu, S.: The bar domain superfamily proteins from subcellular structures to human diseases. Membranes (Basel) **2**, 91–117 (2010). http://www.ncbi.nlm.nih.gov/pubmed/24957964

55. Sánchez-Soriano, N., Tear, G., Whitington, P., Prokop, A.: *Drosophila* as a genetic and cellular model for studies on axonal growth. Neural Dev. **2**, 9 (2007). doi:10.1186/1749-8104-2-9. http://www.ncbi.nlm.nih.gov/entrez/query.fcgi?cmd=Retrieve&db=PubMed&dopt=Citation&list_uids=17475018

56. Sánchez-Soriano, N., Pimentel, C.G., Beaven, R., Haessler, U., Ofner, L., Ballestrem, C., Prokop, A.: *Drosophila* growth cones: a genetically tractable platform for the analysis of axonal growth dynamics. Dev. Neurobiol. **70**, 58–71 (2010). http://www3.interscience.wiley.com/cgi-bin/fulltext/123188795/PDFSTART

57. Vavylonis, D., Yang, Q., O'Shaughnessy, B.: Actin polymerization kinetics, cap structure, and fluctuations. Proc. Natl. Acad. Sci. USA, **102**, 8543–8548 (2005). http://www.ncbi.nlm.nih.gov/pubmed/15939882

58. Wegner, A.: Head to tail polymerization of actin. J. Mol. Biol. **108**, 139–150 (1976). http://www.ncbi.nlm.nih.gov/pubmed/1003481, http://www.sciencedirect.com/science/article/pii/S0022283676801003

59. Wilson, K.G.: Grand challenges to computational science. Futur. Gener. Comput. Syst. **5**, 171–189 (1989). doi:10.1016/0167-739X(89)90038-1

60. Zhuravlev, P.I., Papoian, G.A.: Protein fluxes along the filopodium as a framework for understanding the growth-retraction dynamics: the interplay between diffusion and active transport. Cell Adhes. Migr. **5**, 448–456 (2011)

61. Zhuravlev, P.I., Der, B.S., Papoian, G.A.: Design of active transport must be highly intricate: a possible role of myosin and Ena/VASP for G-actin transport in filopodia. Biophys. J. **98**, 1439–1448 (2010)

62. Zicha, D., Dobbie, I.M., Holt, M.R., Monypenny, J., Soong, D.Y., Gray, C., Dunn, G.A.: Rapid actin transport during cell protrusion. Science **300**, 142–145 (2003)

Chapter 3
Fault Diagnosis with Missing Data Based on Hopfield Neural Networks

Raquelita Torres Cabeza, Egly Barrero Viciedo, Alberto Prieto-Moreno, and Valery Moreno Vega

Abstract Most of the existing artificial neural network models use a significant amount of information for their training. The need for such information could be an inconvenience for its application in fault diagnosis in industrial systems, where the information, due to different factors such as data losses in the data acquisition systems, is scarce or not verified. In this chapter, a diagnostic system based on a Hopfield neural network is proposed to overcome this inconvenience. The proposal is tested using the development and application of methods for the actuator diagnostic in industrial control systems (DAMADICS) benchmark, with successful performance.

Keywords Fault diagnosis • Hopfield neural networks • Industrial processes • Quality of data • Incomplete data • Missing data • DAMADICS

3.1 Introduction

At present, fault diagnosis has acquired great importance in engineering processes due to its potential advantages in the reduction of maintenance and repair costs, the improvement of productivity, and the increase of security and availability of industrial processes [15].

Frequently, when a supervisory control and data acquisition (SCADA) system acquires real-time data from any sub-process in an industrial process, the characteristic data values of these processes might be incomplete and hence lead to loss of information due to several problems; for instance, malfunction of the measurement channel [1, 7, 10, 16]. When these problems occur in an industry, the fault diagnosis methods that rely on historical data lose their effectiveness.

R.T. Cabeza (✉) • E.B. Viciedo • A. Prieto-Moreno • V.M. Vega
Automatic and Computing Department, Instituto Superior Politécnico José Antonio Echeverría (CUJAE), Marianao, La Habana, CP 19390, Cuba
e-mail: rtorresc@electrica.cujae.edu.cu; egly@electrica.cujae.edu.cu; albprieto@electrica.cujae.edu.cu; valery@electrica.cujae.edu.cu

© Springer International Publishing Switzerland 2016   37
A.J. da Silva Neto et al. (eds.), *Mathematical Modeling and Computational Intelligence in Engineering Applications*, DOI 10.1007/978-3-319-38869-4_3

Artificial neural networks (ANNs) are widely used in pattern recognition problems, such as the classification of fault patterns in industrial processes [13]. However, when the information is scarce, its classification hit rate decreases due to the volume of information that ANN training algorithms require for good performance.

The Hopfield network, in its discrete model version, is a neural network architecture that has shown good performance in pattern classification [18]. For the purpose of its training, this network does not require a great volume of information, given the fact that this network operates as associative memory and provides prompt replies thanks to the highly parallel nature of its convergence's process.

This chapter illustrates a fault diagnosis system design that is based on a discrete Hopfield model, the training of which utilized limited amount of information. The results from this design are compared with the probabilistic neural network (PNN), a member of the family of radial basis function neural networks, which needs huge volumes of information for its training.

The structure of the chapter is as follows: Sect. 3.2 discusses the fundamental aspects of Hopfield neural network and the PNN. The DAMADICS benchmark, as well as the design of a fault diagnosis system based on the Hopfield ANN, is also described. At the end of this section, the experiments that were conducted are shown. Section 3.3 discusses the results. Finally, Sect. 3.4 presents the conclusions, as well as recommendations for future studies.

3.2 Materials and Methods

3.2.1 Hopfield and Probabilistic Artificial Neural Networks Models

The ANNs constitute a system of interconnected elements called neurons that accomplish functions such as: learning, memorization, generalization, or abstraction of essential features. Hence, the ANNs represent a distributed structure of parallel processing [3].

3.2.1.1 Hopfield Neural Network

A Hopfield network is designed to associate an input pattern with itself. In other words, if a memory stored pattern is presented at the input, then the network returns the same pattern as its output. If such an input pattern has not been stored, the network evolves to generate the most similar pattern, which allows for the separation and group classification of the different states of the system [3].

The architecture consists of one layer (see Fig. 3.1), where each neuron connects to the rest of them.

The equation that characterizes the network is (3.1) [3].

$$y_i(t) = f(h_i(t)) = f\left(\sum_j W_{ij}X_j(t) - \theta_i\right) \tag{3.1}$$

where:

W_{ij}: synaptic weight between neuron i and neuron j;
X_j: input j of the network;
θ_i: threshold corresponding to neuron i;
h_i: local field corresponding to neuron i;
$f(.)$: activation function (generally sign function);
y_i: output i of the network.

This network learns using a non-supervised Hebbian-type method. This approach finds a set of synaptic weights W in a manner that, for a given data set, the network stores stable states [3, 9].

3.2.1.2 Probabilistic Neural Network

The PNN uses a semi-supervised approach that operates on the basis of the Parzen classifier, as applied to Bayesian statistics.

Given its classifier role, this neural network model has been successfully applied to such areas as pattern recognition and classification [11]. The output layer has one neuron for each class or pattern to be recognized or classified.

PNN architecture is build with four neuron layers (see Fig. 3.1). The first is an input layer consisting of d neurons (dimension of data). The second is a data layer consisting of N neurons (one for each representative vector of the classes). The next is the summation layer formed by k neurons, where k is the number of classes, and the last one is the decision layer that includes one neuron.

Unlike the learning process implemented by most of the ANNs, where adjusted weight and bias parameters are part of the network training, the PNN does not need any weight adjustment. In a PNN, the output patterns are determined by calculating the distance between each input vector and each class-characteristic data, and each input vector is classified on the basis of a weighted distance function, in a manner that the distance between similar training data and their desired output class is shorter than the distance of other training data. In order to achieve the above, during the training phase, the data layer increases its size by adding a neuron for each training data and adjusting the corresponding weight of such added neuron to match the training data vector [8].

When a vector subject to classification is introduced to the network, the second layer computes the distances between the input vector and each class-characteristic data using the function φ_{ij}, as shown in (3.2),

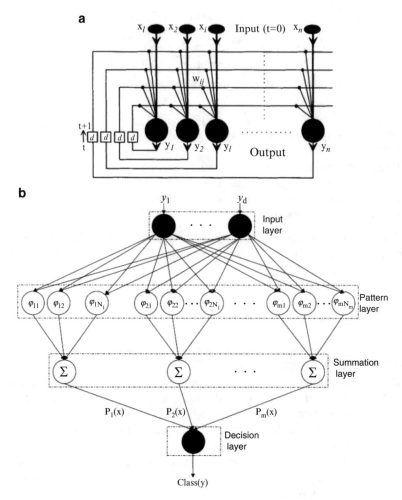

Fig. 3.1 Architectures of Hopfield and PNN neural networks. (**a**) Hopfield neural network; (**b**) Probabilistic neural network

$$\varphi_{ij}(y) = \frac{1}{(2\pi)^{\frac{d}{2}}\sigma^d} \exp\left[-\frac{(y - X_{ij})^T(y - X_{ij})}{2\sigma^2}\right] \tag{3.2}$$

where σ is the dispersion parameter, which takes values between 0 and 1 defined by the designer, and X_{ij} is the j-th data, associated with the class i.

In the third layer, the term $P_i(y)$ indicates the conditional probability or likeness of the input data belonging in the i-th class, as obtained by (3.3),

$$P_i(y) = \frac{1}{N_i} \sum_{j=1}^{N_i} \varphi_{ij}(y) \tag{3.3}$$

where N_i is the number of data of the class i.

Finally, in the fourth layer, i.e., the output layer, the input vector will be assigned to the class with greater likeness, as stated in (3.4) [8, 11],

$$Class(y) - arg\ _i\ max(P_i(y)) \tag{3.4}$$

3.2.2 DAMADICS Benchmarking and Design of Architectures

The benchmark selected to conduct the experiments discussed below is the DAMADICS. Based on an actuator or a final action element that interacts with a process, this benchmark is suitable for testing the performance of fault diagnosis systems. Its different fault modes and both simulated and real data files can be used to test the effectiveness of experimental designs. This benchmark has often been applied to fault diagnosis neural network designed systems [2, 5, 12, 14].

The three actuators included in this benchmark were installed at the Lublin Sugar Factory, Poland. The first two were installed, respectively, at the first and fifth stages of the evaporation station (see Fig. 3.2), for thin and thick juice level control. The third actuator was installed at the steam boiler for water level control of the fourth boiler station (see Fig. 3.3) [2].

The DAMADICS benchmark allows for the simulation of not only general or external faults, but also 19 faults distributed inside the actuator's three main elements: control valve, spring and diaphragm pneumatic servomotor, and positioner [2].

For the purpose of the diagnosis, real data were selected from November 17th 2001, when the actual faults were attributed to an unexpected pressure change across the valve and a flow rate sensor fault.

The first is an incipient, but rapidly developing fault caused by media pump station failure, process disturbance (increased flow output), increased pipes resistance, change of fluid viscosity, external media leakage, etc.

The second is an abrupt fault caused by electronics or wiring failure.

These faults are simulated in all the actuators, but actuator 2 operates within the flow control loop. In this case, the abrupt fault was introduced only in the control loop. The flow signal available in the DAMADICS data file is fault free (see Fig. 3.4).

The variables of interest are

- Pressure valve inlet **P1**
- Pressure valve outlet **P2**
- Process value (Flow) **PV**

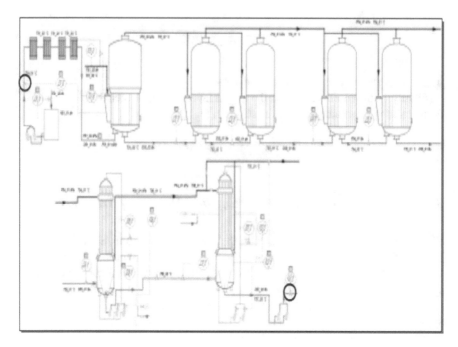

Fig. 3.2 Evaporation station diagram

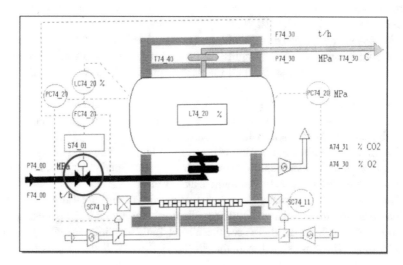

Fig. 3.3 Steam boiler diagram

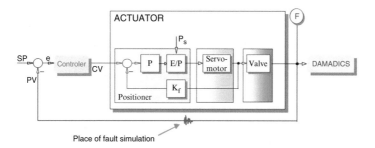

Fig. 3.4 Scheme of the flow control loop of thick juice outflow from the fifth stage of evaporation station

- Controller output (control value) **CV**
- Servomotor rod displacement **X**

In this chapter, the aforementioned two faults, as well as the state of normal operation, are diagnosed, for a total of three classes.

3.2.2.1   Hopfield Design and Configuration

Generally, when ANNs are applied to fault diagnosis, the number of neurons in the input layer coincides with the number of variables that are measured in the process. In this case, Hopfield is a single-layer architecture with five neurons in its input layer because the data set has five main components (variables). The data processing occurs in its single layer, and the recovered pattern is supplied as output. Since the recovered pattern has the same dimension as the input vector, this layer is also composed of five neurons.

The network only produces a data set to its output; consequently, a subsequent classification mechanism is required for the classification of data into their respective classes. In order to execute this mechanism, all the outputs from the network must be obtained and the following algorithm implemented:

1. The average vector of the input data that represents each class is determined.
2. The Euclidean distance between each NN output data and the average vector is calculated.
3. The smallest distance is determined.
4. Each class has an output pattern that will reflect the smallest calculated Euclidean distance from the average vector of the input data.

The Euclidean distance was chosen for its widespread application in NN designs and its short computation time [3].

3.2.2.2 PNN Design and Configuration

For the purpose of estimating the probability density function as the standard kernel, the Gaussian kernel function was chosen [8, 11].

The dispersion parameter takes a value between 0 and 1, which is why $\sigma = 0.5$ is selected.

As the number of neurons in the input layer matches the number of variables measured in the process, the PNN, as is the case with Hopfield, has five neurons in its input layer, its data layer contains as many neurons as training data, its summation layer includes three neurons representing three classes, and its decision layer consists of a single neuron reflecting the class being associated with each input data.

3.2.3 Design of the Experiments

Consistently with the proposed ANN models, a set of experiments were conducted. Such experiments relied on 100 input data, including 46 from the normal operation state, 15 associated with the incipient fault, and 39 pertaining to the abrupt fault.

Two types of experiments were designed, and the cross validation method was applied therein [17] by dividing the data of each class into $k = 5$ partitions:

- Training the networks with $k - 1$ partitions and 1 partition to validate.
- Training the networks with 1 partition and $k - 1$ partitions to validate.

These two experiments help assess the effectiveness of the ANN when the amount of training data decreases (experiment 1 uses more data sets than experiment 2); in other words, a scenario where the information is scarce or incomplete.

3.3 Results and Discussion

After completion of the cross validation method, five classifying errors were obtained: one for each partition. Then, the average number of classification errors was computed for use as the classification error produced by the ANNs.

Tables 3.1 and 3.2 show the results of the experiments.

Table 3.1 Rate of classification error using cross validation method

Architectures	Rate of classification error (%)
Hopfield	11
PNN	8

Table 3.2 Rate of classification error using cross validation method with few training data

Architectures	Rate of classification error (%)
Hopfield	10
PNN	18.25

A Wilcoxon statistic test was conducted to check for significant differences, if any, between the results obtained using the Hopfield network and the PNN. The Wilcoxon statistic test is a simple, safe, and robust non-parametric procedure. It permits to draw statistical paired comparisons in order to determine significant behavioral differences between two algorithms [4, 6]. When the test was applied to the first experiment, its results indicated the inexistence of significant differences among the classification errors made by the ANNs with a 95 % confidence level and a p-value equal to 0.797248. In the second experiment, the same test found, with a 95 % confidence level and a p-value of 0.0202405, that significant statistical differences existed between the two classification errors.

Specifically, these Wilcoxon test results show that the application of the PNN is not the same as the Hopfield network. These results also confirm that the latter is a better option (fewer classification errors).

3.4   Conclusions

This project deals with a neural network based fault diagnosis in systems where loss of information might have occurred. Actual experiments have statistically shown that the Hopfield neural network offers better fault-classification results than the PNN; especially, where the available data in the training process is scant. Furthermore, the classification errors fall within a 10 % range, which is acceptable. Therefore, the Hopfield network appears to be a good candidate for fault diagnosis systems; in particular, where the available information is scarce.

References

1. Baraldi, P., Maio, F.D., Genini, D., Zio, E.: Reconstruction of missing data in multidimensional time series by fuzzy similarity. Appl. Soft Comput. **26**, 1–9 (2015)
2. Bartys, M., Patton, R., Syfert, M., de las Heras, S., Quevedo, J.: Introduction to the DAMADICS actuator FDI benchmark study. Control. Eng. Pract. **14**, 577–596 (2006)
3. del Brio, B., Molina, A.: Redes Neuronales y Sistemas Difusos, Departamento de Ingeniería Electrínica y Comunicaciones. Universidad de Zaragoza, Zaragoza, España. Marzo de, 2nd edn. RA-MA (2001)
4. Capanu, M., Jones, G., Randles, R.: Testing for preference using a sum of Wilcoxon signed rank statistics. Comput. Stat. Data Anal. **51**, 793–796 (2006)
5. Chopra, T., Vajpai, J.: Classification of faults in DAMADICS benchmark process control system using self organizing maps. Int. J. Soft Comput. Eng. **1**(3), 85–90 (2011)

6. Dewan, I., Rao, B.P.: Wilcoxon-signed rank test for associated sequences. Stat. Probab. Lett. **71**, 131–142 (2005)
7. He, X., Wang, Z., Zhou, D.: Robust fault detection for networked systems with communication delay and data missing. Automatica **45**, 2634–2639 (2009)
8. Khashei, M., Bijari, M.: Hybridization of the probabilistic neural networks with feed-forward neural networks for forecasting. Eng. Appl. Artif. Intell. **25**, 1277–1288 (2012)
9. Kumar, S., Singh, M.: Study of Hopfield neural network with sub-optimal and random GA for pattern recalling of English characters. Appl. Soft Comput. **12**, 2593–2600 (2012)
10. Markey, M., Tourassi, G., Margolis, M., DeLong, D.: Impact of missing data in evaluating artificial neural networks trained on complete data. Comput. Biol. Med. **36**, 516–525 (2006)
11. Miguez, R., Georgiopoulos, M., Kaylani, A.: G-PNN: A genetically engineered probabilistic neural network. Nonlinear Anal. **73**, 1783–1791 (2010)
12. Patan, K.: Artificial Neural Networks for the Modelling and Fault Diagnosis of Technical Processes. Springer, Berlin (2008)
13. Simani, S., Fantuzzi, C., Patton, R.J.: Model-Based Fault Diagnosis in Dynamic Systems Using Identification Techniques. Springer, London (2002)
14. Uppal, F., Patton, R., Witczak, M.: A neuro-fuzzy multiple-model observer approach to robust fault diagnosis based on the DAMADICS benchmark problem. Control Eng. Pract. **14**, 699–717 (2006)
15. Venkatasubramanian, V., Rengaswamy, R., Yin, K., Kavuri, S.N.: A review of process fault detection and diagnosis part i: Quantitative model-based methods. Comput. Chem. Eng. **27**(3), 293–311 (2003)
16. Wu, X., Wang, Y., Mao, J., Du, Z., Li, C.: Multi-step prediction of time series with random missing data. Appl. Math. Model. **38**(14), 3512–3522 (2014)
17. Yaghini, M., Khoshraftar, M., Fallahi, M.: A hybrid algorithm for artificial neural network training. Eng. Appl. Artif. Intell. **26**, 293–301 (2013)
18. Zhiqiang, L., Li, Z., Xue, L., Jie, C.: Evaluation method about bus scheduling based on discrete Hopfield neural network. J. Transp. Syst. Eng. Inf. Technol. **11**(2), 77–83 (2011)

Chapter 4
Diagnosing Time-Dependent Incipient Faults

Lídice Camps-Echevarría, Orestes Llanes-Santiago, Haroldo Fraga de Campos Velho, and Antônio José da Silva Neto

Abstract This chapter focuses on a formulation for fault diagnosis (FDI) using an inverse problem methodology. It has been shown that this approach allows for diagnoses with adequate balance between robustness and sensitivity. The main contribution of this chapter is the expansion of this approach to include the diagnosis of time-dependent incipient faults. The FDI inverse problem is formulated as an optimization problem that is then solved with two metaheuristics: Differential Evolution and its variation Differential Evolution with Particle Collision. The proposed methodology is tested using simulated data from the Two Tanks system, which is recognized as benchmark for control and diagnosis. The results indicate that this proposal is suitable for the aforementioned diagnosis.

Keywords Model based fault diagnosis • Incipient faults • Inverse problems • Differential Evolution algorithm • Particle Collision Algorithm • Robustness • Sensitivity

L. Camps-Echevarría
CEMAT, Instituto Superior Politécnico José Antonio Echeverría (CUJAE),
Marianao, La Habana, CP 19390, Cuba
e-mail: camps.lidice@gmail.com

O. Llanes-Santiago (✉)
Automatic and Computing Department, Instituto Superior Politécnico José Antonio Echeverría (CUJAE), Marianao, La Habana, CP 19390, Cuba
e-mail: orestes@tesla.cujae.edu.cu

H.F. de Campos Velho
INPE, Av. dos Astronautas 1758, São José dos Campos, SP, CEP 12.227-010, Brazil
e-mail: haroldo@lac.inpe.br

A.J. Silva Neto
Mechanical Engineering and Energy Department, Polytechnic Institute, IPRJ-UERJ,
Rua Bonfim, No. 25, Nova Friburgo, RJ, CEP 28625-570, Brazil
e-mail: ajsneto@iprj.uerj.br

© Springer International Publishing Switzerland 2016
A.J. da Silva Neto et al. (eds.), *Mathematical Modeling and Computational Intelligence in Engineering Applications*, DOI 10.1007/978-3-319-38869-4_4

47

4.1 Introduction

Fault diagnosis, FDI, involves detection, isolation, and identification of faults [10]. This is an important task for improved reliability and safety in the industrial sector [10, 23, 24].

FDI methods are expected to guarantee fast detection of faults, while rejecting false alarms associated with certain factors, such as uncertainties in measurements, external disturbances, and/or spurious signals. The above led to the need for FDI methods that handle an adequate balance between sensitivity and robustness [10, 23, 24].

For practical applications of FDI methods, another desirable characteristic is added: a reasonable computational cost that allows for diagnosing online processes in almost real time [7, 10, 23, 24, 32]. Furthermore, the need to develop new FDI methods that handle this balance appropriately is still recognized [7, 23, 24, 32].

Within FDI methods, the analytical model-based methods are found [29]. The main approaches within the model-based methods are: observers, parity space, and parameter estimation [7, 9, 10, 24]. The main limitations of each of these approaches are explained in [7, 31, 32].

A closer look to FDI model-based approaches show that they share a common structure: knowledge of the system's outputs and inputs, as well as a model that describes the system, and a capacity to determine the faults affecting the system [7, 24, 32]. This structure coincides with the inverse problems structure [22].

It has been recently shown that formulating fault diagnosis by an inverse problem methodology helps obtain an appropriate balance between robustness and sensitivity [2, 4, 33]. In all these papers, FDI inverse problem is established as an optimization problem that is solved with metaheuristics. As a first approximation, these works assume that faults keep a constant magnitude over time [2, 4, 33].

This chapter also focuses on a formulation of fault diagnosis by an inverse problem methodology. The main contribution of this chapter is the expansion of this method to include the diagnosis of incipient faults that are time dependent. The application of results from diagnosis, namely results related with structural detectability and structural separability of faults [11], helps obtain information concerning the inverse problem under study [4].

Considering the reported applications of metaheuristics to both FDI [2, 13, 16, 21, 30, 31, 33] and parameter estimation inverse problems [19, 22, 27], this paper also seeks to solve the optimization problem using metaheuristics. In this case, two metaheuristics are applied: Differential Evolution, DE [25], and its modified version designated as Differential Evolution with Particle Collision, DEwPC [5]. This selection is based on their previous applications in FDI that showed good results for a robust and sensitive diagnosis [3, 5]. The proposed methodology is tested using simulated data from the Two Tanks system, which is recognized as benchmark for control and diagnosis [14]. The results show the suitability of this proposal.

The remaining content of this chapter is organized as follows. In Sect. 4.2, FDI is presented as an inverse problem. The methods DE and DEwPC are briefly explained in Sect. 4.3. Section 4.4 discusses the case of study and its simulations. The following Sects. 4.5 and 4.6 present the experimental methodology and results, respectively. In Sect. 4.7, some concluding comments and remarks are presented.

4.2 Fault Diagnosis as an Inverse Problem

The parameter estimation approach is a distinctive method within FDI model-based approaches [10, 24]. The FDI methods that are based on model parameters involve online parameter estimation methods. They require knowledge of both the input vector $u(t)$ and the output vector $y(t)$, as well as a basic model structure that describes the system [7, 10, 24].

In this chapter, the following nonlinear state space model is considered:

$$\dot{x}(t) = g(x(t), u(t), \theta)$$
$$y(t) = h(x(t), u(t)) \tag{4.1}$$
$$x(t_0) = x_0$$

where $x(t) \in \mathbf{R}^n$ is the state variables vector; x_0 is the initial state; $\theta \in \mathbf{R}^l$ is the parameter vector of the model, and $t \in [t_0, t_f]$. The input $u(t) \in \mathbf{R}^p$ and the output $y(t) \in \mathbf{R}^m$ are measured with sensors.

The faults affecting system may change the parameter values in vector θ [7, 10]. The main disadvantage of this approach is that the model's parameters should have physical meaning; i.e., they should match the parameters of the system [7, 10]. Furthermore, fault isolation may become extremely difficult because the model parameters fail to match those of the system [10, 32] on a one-to-one basis.

Alternatively, a model that directly includes the dynamics of the faults, by means of a fault vector f [7, 9], can be assessed as follows:

$$\dot{x}(t) = g(x(t), u(t), f)$$
$$y(t) = h(x(t), u(t), f) \tag{4.2}$$
$$x(t_0) = x_0$$

The fault vector $f = [f_a f_p f_s]^T \in \mathbf{R}^{p+q+m}$, where $f_a \in \mathbf{R}^p$; $f_p \in \mathbf{R}^q$, and $f_s \in \mathbf{R}^m$ are faults affecting the actuator, process, and sensors, respectively [7, 9]. These three parts establish the level of abstraction for diagnosis [24]. In such cases, diagnosis can be directly obtained from estimating the fault vector $\hat{f}$. This is an inverse problem of parameter estimation that can be formulated as an optimization problem:

$$\min F(\hat{f}) = \sum_{t=1}^{S} \left[y(f, t) - \hat{y}(\hat{f}, u(t), t) \right]^2 \tag{4.3}$$
$$s.t.\ f_{min} \leq \hat{f} \leq f_{max}$$

where S is the total number of sampling instants, $\hat{y}(\hat{f}, u(t), t)$ is the estimated vector output at each instant of time, which is obtained from the model (4.2) using the measurements of the input $u(t)$, and $y(f, t)$ is the output vector, which is measured by sensors at the same time instant t.

In recent works, this approach has been applied to FDI [2, 4, 33]. In these previous works, the faults were assumed constant throughout the process. The main contribution of this chapter, and its difference from these previous works, is that faults are considered to be time dependent. The proposal herein intends to include the faulty conditions that affect systems in practical situations, where faults need to be modeled as ramp functions in the following form:

$$f_i(t) = m_i t \tag{4.4}$$

The optimization problem becomes

$$\min F(\hat{m}) = \sum_{t=1}^{S} [y(m, t) - \hat{y}(\hat{m}, u(t), t)]^2 \tag{4.5}$$
$$s.t. \ m_{\min} \leq \hat{m} \leq m_{\max}$$

where $m_{\min}$, $m_{\max}$ are functions of $f_{\min}$, $f_{\max}$, respectively. This problem is similar to other parameter estimation inverse problems. Furthermore, in light of reported applications of metaheuristics to both FDI [2, 13, 16, 21, 30, 31, 33] and parameter estimation inverse problems [19, 22, 27], this proposal also seeks to solve the problem given in (4.4) and (4.5) by means of metaheuristics.

4.2.1   Structural Analysis

For the purpose of collecting prior information about the uniqueness or non-uniqueness of the set of fault vectors that may justify the observed behavior of the system, some results arising from sensor placement for faults detectability and separability are applied [11, 12]. These results are based on a structural representation of the model and the Dulmage–Mendelsohn decomposition [8]. The work of [4] shows how the above can be understood as an alternative to the sensitivity analysis for parameter estimation inverse problems, when the model of the system is represented by ordinary differential equations.

Let us assume $E = \{e_1, e_2, \ldots, e_h\}$ as the set of equations in a model, and $X = \{x_1, x_2, \ldots, x_k\}$ as the set of unknown variables in the same model. Let us also assume that each fault f_l affects only one equation that is denoted by e_{f_l}. The bi-adjacency matrix in M for the bipartite graph $G_b = (E, X)$ that represents the model is defined as in $M = \{m_{ij}\}$ with:

$$m_{ij} = \begin{cases} 1 & x_j \vee \dot{x}_j \in e_i \\ 0 & otherwise \end{cases} \tag{4.6}$$

Based on a bi-adjacency matrix of a system, the structurally detectable faults and the structural separable faults have been defined [11].

Definition 4.1. Let us suppose that the adjacency matrix in M of a model of ordinary differential equations and a fault f_l that affects the equation e_{f_l} of this model are known. Then, the following definition applies:

Fault f_l is structurally detectable if $e_{f_l} \in M^+$.

Definition 4.2. Let us suppose that the adjacency matrix in M of a model of ordinary differential equations and two faults f_i and f_j that affect the equations e_{f_i} and e_{f_j} of this model, respectively, are known. Then, the following definition applies:

Fault f_i is structurally separable from f_j if $e_{f_i} \in M^+$, where M^+ denotes the overdetermined part of a model M, which can be obtained with the Dulmage–Mendelsohn decomposition [8, 12].

4.3 Differential Evolution and Differential Evolution with Particle Collision

4.3.1 Differential Evolution

Differential Evolution, DE, was proposed in 1995 for optimization problems [25]. Some of the most important advantages of DE are: simple structure, simple computational implementation, and speed and robustness [22, 25]. DE is based on three operators: *Mutation*, *Crossover*, and *Selection* [22, 25]. These operators are based on vector operations, as their main difference from genetic algorithms [22]. At each iteration *Iter* the algorithm generates a new population of Z feasible solutions $X_{\text{Iter}}^1, X_{\text{Iter}}^2 \ldots X_{\text{Iter}}^Z$ with the application of its three operators on the current population. This mechanism can be summarized with the notation:

$$DE/X_{\text{Iter}-1}^\delta/\gamma/\lambda \tag{4.7}$$

where γ indicates the number of pairs of solutions from the current population to be used for perturbing the present solution $X_{\text{Iter}-1}^\delta$; λ represents a distribution function to be used during *Crossover*. This chapter applied the scheme $DE/X^{\text{best}}/2/bin$, where *bin* is a notation for a binomial distribution function and *Mutation* is described by:

$$X_{\text{Iter}}^z = X^{\text{best}} + C_{\text{scal}} \left(X_{\text{Iter}-1}^{\alpha 1} - X_{\text{Iter}-1}^{\alpha 3} + X_{\text{Iter}-1}^{\alpha 2} - X_{\text{Iter}-1}^{\alpha 4} \right) \tag{4.8}$$

where $X^{\text{best}}, X_{\text{Iter}-1}^{\alpha 1}, X_{\text{Iter}-1}^{\alpha 2}, X_{\text{Iter}-1}^{\alpha 3}, X_{\text{Iter}-1}^{\alpha 4} \in \mathbf{R}^n$ are solutions from the current population and C_{scal} is an algorithm's parameter, called *Scaling factor*. *Crossover* and *Selection* operators can be described as:

• *Crossover*

$$\hat{x}_{(\text{Iter})n}^z = \begin{cases} \hat{x}_{(\text{Iter})n}^z & \text{if } q_{\text{rand}} \leq C_{\text{cross}} \\ \hat{x}_{(\text{Iter})n}^\delta & \text{otherwise} \end{cases} \tag{4.9}$$

Fig. 4.1 Algorithm for the
Differential Evolution method
(DE)

> **Input:** $Z, \textit{MaxIter}, C_{\textit{scal}}, C_{\textit{cross}}$
> **Output:** $\mathbf{X}^{\textit{best}}$
> 1: Generate an initial population of Z solutions
> 2: Select best solution $X^{\textit{best}}$
> 3: **for** $\textit{Iter} \leftarrow 1, \textit{MaxIter}$ **do**
> 4:   Apply *Mutation*
> 5:   Apply *Crossover*
> 6:   Apply *Selection*
> 7:   Update $X^{\textit{best}}$
> 8:   Verify stopping criteria
> 9: **end for**
> 10: Solution: $\mathbf{X}^{\textit{best}}$

where $\hat{x}^z_{(\textit{Iter})n}$ are components from vector $\hat{\mathbf{X}}^z_{\textit{Iter}}$; $0 \leq C_{\textit{cross}} \leq 1$ is another algorithm's parameter: *crossover factor*; and $q_{\textit{rand}}$ is a random number that is generated by means of the distribution represented by λ.

- *Selection*

Vector $\mathbf{X}^z_{\textit{Iter}}$ to be part of a new population is selected using the following rule:

$$\mathbf{X}^z_{\textit{Iter}} = \begin{cases} \hat{\mathbf{X}}^z_{\textit{Iter}} & \textit{if } F(\hat{\mathbf{X}}^z_{\textit{Iter}}) \leq F(\mathbf{X}^\delta_{\textit{Iter}-1}) \\ \mathbf{X}^\delta_{\textit{Iter}-1} & \textit{otherwise} \end{cases} \qquad (4.10)$$

A general description of the algorithm for DE optimization algorithm is shown in Fig. 4.1.

The most successful versions of *DE* have focused on variations of the *Mutation* operator and the self adaptation of the parameters $C_{\textit{cross}}$ and $C_{\textit{scal}}$ [1, 6, 17, 18, 28, 34, 35].

4.3.2 Differential Evolution with Particle Collision

The recently proposed algorithm Differential Evolution with Particle Collision, *DEwPC* [3, 5], is intended to improve the performance of DE based on the incorporation of some ideas from the Particle Collision Algorithm, *PCA* [15, 19, 20], in order to improve its capacity for avoiding local optima. *DEwPC* keeps the same structure of the operators *Mutation* and *Crossover* in *DE*, while introduces a modification in the *Selection* operator [3, 5]. This modification adds a new parameter *MaxIter_c*. The new *Selection* operator takes the ideas of *Absorption* and *Scattering* from *PCA*. The adaptation of this operator to *DEwPC* has been called *Selection with Absorption-Scattering with probability* and this adaptation can be established as [3, 5]:

Fig. 4.2 *Absorption* Operator

> **Input: $\hat{\mathbf{X}}_{Iter}$**
> **Output: $\mathbf{X}_{Iter}$**
>   1: $\mathbf{X}_{Iter} = \hat{\mathbf{X}}_{Iter}$
>   2: Small Search $(\mathbf{X}_{Iter})$

Fig. 4.3 *Scattering with Probability* Operator

> **Input: $\hat{\mathbf{X}}_{Iter}, F(\mathbf{X}^{best})$**
> **Output: $\mathbf{X}_{Iter}$**
>   1: Compute $F(\hat{\mathbf{X}}_{Iter})$
>   2: Compute $p_{r(Iter)} = 1 - \frac{F(\mathbf{X}^{best})}{F(\hat{\mathbf{X}}_{Iter})}$
>   3: Generate a random number q
>   4: **if** $q < p_{r(Iter)}$ **then**
>   5: $\mathbf{X}_{Iter} = \hat{\mathbf{X}}_{Iter}$
>   6:   Search $(\mathbf{X}_{Iter})$
>   7: **else**
>   8: $\mathbf{X}_{Iter} = \mathbf{X}_{Iter-1}$
>   9: **end if**

Selection with Absorption-Scattering with probability (Nsp):

- If $F(\hat{\mathbf{X}}^z_{Iter}) \leq F(\mathbf{X}^{\delta}_{Iter-1})$, then the operator *Absorption* is applied to $\hat{\mathbf{X}}^z_{Iter}$.
- If $F(\hat{\mathbf{X}}^z_{Iter}) > F(\mathbf{X}^{\delta}_{Iter-1})$, then the operator *Scattering with probability* is applied to $\hat{\mathbf{X}}^z_{Iter}$.

The *Absorption* and *Scattering with probability* operators are represented in Figs. 4.2 and 4.3, respectively.

Figure 4.4 shows the algorithm for *DEwPC*. *Small Search* operator indicates a small stochastic perturbation around a solution [19, 20, 22].

4.4 Two Tanks System

The Two Tanks system is a simplified version of the Three Tanks system [14]. Both are benchmarks for control and diagnosis. The system is formed by two tanks of liquid that can be filled with two similar and independent pumps acting on tanks 1 and 2, respectively. Both tanks have the same cross section $S_1 = S_2 = 2.54$ m^2. The pumps deliver flow rate q_1 in tank 1 and q_2 in tank 2. The tanks are interconnected through underflow pipes, see Fig. 4.5. All the pipes have the same cross section $S_p = 0.1$ m^2. The liquid levels L_1 and L_2 at tanks 1 and 2, respectively, are outputs of the system. The control variables q_1 and q_2 are chosen to control the levels of tank 1 and tank 2 ($\tilde{L}_1 = 4.0$ m and $\tilde{L}_2 = 3.0$ m).

Each tank is leaking due to a hole in its bottom. These leaks are regarded as faults affecting the proper behavior of the system. Every leak allows for a liquid outflow rate called $f_{p(1)}(t)$ and $f_{p(2)}(t)$ in tanks 1 and 2, respectively. It is assumed that these

Fig. 4.4 Algorithm for
DEwPC

Input: Z, *MaxIter*, C_{scal}, C_{cross}, *MaxIter$_c$*
Output: $\mathbf{X}^{best}$
1: Generate an initial population of Z solutions
2: Select best solution X^{best}
3: **for** *Iter* $\leftarrow 1$, *MaxIter* **do**
4:　　Apply *Mutation*
5:　　Apply *Crossover*
6:　　**for** $j \leftarrow 1, Z$ **do**
7:　　　　**if** *rand* < 0.7 **then**
8:　　　　　　Apply *Nsp* to $\hat{\mathbf{X}}^{(j)}_{Iter}$
9:　　　　**else**
10:　　　　　Apply *Selection* to $\hat{\mathbf{X}}^{(j)}_{Iter}$
11:　　　**end if**
12:　　**end for**
13:　　Update X^{best}
14:　　Verify stopping criteria
15: **end for**
16: Solution: $\mathbf{X}^{best}$

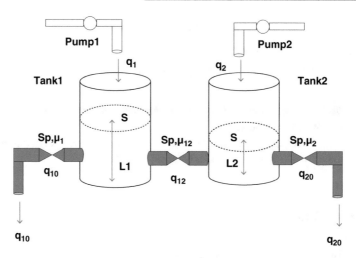

Fig. 4.5 Two tanks system (adapted from [14])

leaks do not release more than 1 m^3 of liquid per second. This assumption can be
described with the following restrictions:

$$f_{p(1)}, f_{p(2)} \in \mathbf{R} : 0 \leq f_{p(1)}, f_{p(2)} \leq 1 \text{ m}^3/\text{s} \quad (4.11)$$

The system's model is derived from the application of fundamental laws.

$$\dot{L}_1 = \frac{q_1}{S_1} - \frac{C_1}{S_1}\sqrt{L_1} - \frac{C_3}{S_1}\sqrt{|L_1 - L_2|}sign(L_1 - L_2) - \frac{f_{p(1)}}{S_1} \quad (4.12)$$

$$\dot{L}_2 = \frac{q_2}{S_2} - \frac{C_2}{S_2}\sqrt{L_2} - \frac{C_3}{S_2}\sqrt{|L_1 - L_2|}\text{sign}(L_1 - L_2) - \frac{f_{p(2)}}{S_2}$$

$$y_1 = L_1$$

$$y_2 = L_2$$

The model represented in (4.12) is nonlinear; $C_i = \mu_i S_p \sqrt{2g}$, with $i = 1, 2$, and $C_3 = \mu_{12} S_p \sqrt{2g}$ are set as $C_1 = C_2 = C_3 = 0.3028 \text{ m}^{5/2}/\text{s}$. The first two equations of the model represent the conservation law for Tanks 1 and 2, respectively. The elements $\frac{f_{p(1)}}{S_1}$ and $\frac{f_{p(2)}}{S_2}$ are related with the liquid that leaks from Tanks 1 and 2, respectively, that is why they appear in their respective equations. A Proportional-Integral-Derivative controller was designed with parameters $K_p = [12\ 14]$, $K_i = [1.15\ 0.3]$, and $K_d = [1.0\ 1.5]$.

The leaks at each tank are considered to be incipient, but growing in dependence of time. These changes were modeled on the basis of a ramp function:

$$f_{p(1)} = m_1 t\ ,\ f_{p(2)} = m_2 t\ ,\ \text{with}\ 0 \le m_1 \le \frac{1}{t_f}\ ,\ 0 \le m_2 \le \frac{1}{t_f} \qquad (4.13)$$

The inverse problem of estimating the fault vector is formulated as an optimization problem:

$$\min \|F(\hat{m}_1, \hat{m}_2)\|_\infty = \left\|\sum_{t=1}^S \left(L(t, m_1, m_2) - \hat{L}(t, q_1, q_1\hat{m}_1, \hat{m}_2)\right)^2\right\|_\infty$$

$$\text{s.t.} \qquad\qquad 0 \le \hat{m}_1 \le \tfrac{1}{t_f} \qquad\qquad (4.14)$$

$$0 \le \hat{m}_2 \le \tfrac{1}{t_f}$$

where $L^T = (L_1\ L_2) \in \mathbf{R}^2$ are the liquid levels obtained by direct measurement of the output, $\hat{L}^T = (\hat{L}_1\ \hat{L}_2) \in \mathbf{R}^2$ are the estimated liquid levels obtained using the model (4.12) and $\|g\|_\infty = \max_i g_i$. This optimization problem is solved with the application of the stochastic algorithms *DE* and *DEwPC*.

4.4.1 Structural Analysis

A model that describes the two tanks system is represented by the four equations in (4.12). Each fault $f_{p(1)}$ and $f_{p(2)}$ affects only one equation (first and second equation of the model (4.12), respectively):

$$e_{f_{p(1)}} = e_1\ ,\ e_{f_{p(2)}} = e_2 \qquad\qquad (4.15)$$

Considering the bipartite graph of the model represented in (4.12): $G_b = (E, X)$, where $\mathbf{E} = \{e_1, e_2, e_3, e_4\}$ and $\mathbf{X} = \{L_1, L_2\}$, its bi-adjacency matrix is

$$
\begin{array}{c|cc}
\text{eq/var} & L_1 & L_2 \\
\hline
e_1 & 1 & 1 \\
e_2 & 1 & 1 \\
e_3 & 1 & 0 \\
e_4 & 0 & 1
\end{array}
\tag{4.16}
$$

After the Dulmage–Mendelsohn decomposition of this model, which is obtained using Matlab® and its function *dmperm*, it can be established that $M^+ = M$. Therefore, it can be concluded that both faults are detectable. For a separability analysis, there is need to study one case: $f_{p(1)}$ separable from $f_{p(2)}$. Taking into consideration Definition 4.2 , it is confirmed that $f_{p(1)}$ is separable from $f_{p(2)}$ when:

$$
e_{f_{(p1)}} \in (M \setminus e_{f_{(p2)}})^+ \leftrightarrow e_{f_{(p1)}} \in (M \setminus e_2)^+
\tag{4.17}
$$

The bi-adjacency matrix for $M \setminus e_2$ is

$$
\begin{array}{c|cc}
\text{eq/var} & L_1 & L_2 \\
\hline
e_1 & 1 & 1 \\
e_3 & 1 & 0 \\
e_4 & 0 & 1
\end{array}
\tag{4.18}
$$

After the Dulmage–Mendelsohn decomposition has been obtained, it is established that $(M \setminus e_2)^+ = M \setminus e_2$. Therefore, $e_1 = e_{f_{(p1)}} \in (M \setminus e_2)^+$, which means that $f_{p(1)}$ is separable from $f_{p(2)}$.

These structural separability and detectability results [4, 11], indicate that, in this case the two faults can be estimated at the same time with the measurements available for the system under analysis.

4.5 Experimental Methodology

For the purpose of assessing the feasibility of the inverse problem methodology for diagnosing time-dependent incipient faults, three criteria were considered: quality of estimations, robustness, and computational cost.

Faults dynamics are described by Eqs. (4.11)–(4.13). Table 4.1 shows eight different cases to be diagnosed during the experiments. Cases 1, 3, 5, 7 and Cases 2, 4, 6, 8 represent the same fault situations, but with different noise levels in measurement for L_1 and L_2 (up to 2, 5, 8, and 10 %) in order to evaluate robustness of the methodology employed. The first group centers on a fault in tank 1 only, while the second group has leaks in both tanks. These faulty situations are intended to simulate incipient faults; in other words, by the end of the sampling time $t_f = 100$ s, the value of a fault is small, and its effect on the outputs of the test system may be masked by noise's effect.

Table 4.1 Cases considered
in numerical experiments

Case	m_1	m_2	Noise level %
1	0.006	0	2
2	0.006	0.004	2
3	0.006	0	5
4	0.006	0.004	5
5	0.006	0	8
6	0.006	0.004	8
7	0.006	0	10
8	0.006	0.004	10

The computational cost is compared using the number of objective function evaluations as the criterion. This point is important because it determines the diagnosis time, which is a requirement for online processes. For this analysis, the algorithms are executed under two stopping criteria: maximum number of iterations and estimation errors. For each case, 25 independent runs of each algorithm were executed. Two indicators were computed: Success Rate (SR), and Success Performance (SP) [26]. The former indicates the percentage of successful runs $SR = \frac{\#SuccessfulRuns}{TotalofRuns}.100$, in which a successful run is a run that has finished because the estimation error for both parameters has reached an acceptable value (in this case acceptable values of estimation error are $e \leq 2\%R$, where R is the true value of the parameter). Parameter SP is computed by means of $SP = \frac{\overline{Eval(EE)}}{SR}$ being $\overline{Eval}(EE)$ the average number of objective function evaluations for the successful runs [26]. Parameter SP is associated with the number of objective function evaluations needed for a successful run.

Implementation of DE It is based on the algorithm in Fig. 4.1. The parameters values are: $Z = 20$, $C_{cross} = 0.9$, and $C_{scal} = 0.5$. The stopping criteria are $MaxIter = 100$ or $F(\hat{m}_1, \hat{m}_2) = M(noise)$.

Implementation of DEwPC It is based on the algorithm in Fig. 4.4. The parameters values considered in this chapter are: $Z = 10$, $C_{cross} = 0.9$, $C_{scal} = 0.5$, $MaxIter_c = 5$. The stopping criteria are $MaxIter = 100$ or $F(\hat{m}_1, \hat{m}_2) = M(noise)$.

4.6 Results

Figure 4.6 shows the average values of the relative estimation error obtained with each algorithm for Cases 1, 3, 5, and 7. The results indicate that as the noise level increases, the estimated relative error level rises in both algorithms DE and $DEwPC$. For this first part of the experiments, a successful run was defined as one in which the values for the error in both estimations were $e \leq 2\%R$, being R the true value of the parameter (and this value was actually obtained from the numerical experiments that were conducted). When the aforementioned error criterion was not satisfied, a

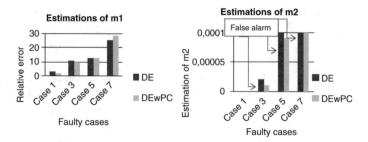

Fig. 4.6 Average relative estimation errors for Cases 1, 3, 5, and 7

Fig. 4.7 Values of the
objective function at different
noise levels and correct
parameters estimation

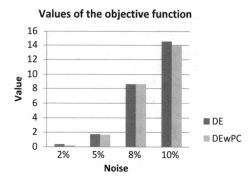

false alarm was triggered. As illustrated in Fig. 4.6 a noise level increase leads to false alarms due to an overestimation of the fault parameters. In all tested cases, the algorithms reached *MaxIter*. It must be recalled that the objective function depends on the estimated values for the fault parameters. Therefore, a decision was made to study the values of the objective function for different noise levels when the parameters estimations are correct, and to try to reduce the number of false alarms based on the knowledge of the noise level affecting the system.

Figure 4.7 shows the values of the objective function for different noise levels. It was established that no faults were affecting the system. Consequently, it was possible to change the stopping criterion $F(\hat{m}_1, \hat{m}_2)$ to reflect its dependence on the noise affecting the system $F(\hat{m}_1, \hat{m}_2) = M(noise)$. For the above, prior information concerning the noise level affecting the measurements is desirable.

After the stopping criterion $F(\hat{m}_1, \hat{m}_2) = M(noise)$ was adjusted, the experiments were repeated. The results are shown in Figs. 4.8 and 4.9. In this case, the number of false alarms dropped; a reflection of increased robustness. For a noise level up to 8 %, the relative error is kept under 10 %. For a noise level up to 10 %, the relative error is kept under 13 % for DE and near to 10 % for *DEwPC*. These results indicate that for highly noisy environments, the *DEwPC* estimations are more accurate, which means that the diagnosis is more robust.

In regard to the computational cost, Table 4.2 shows the indicators *SR* and *SP*. In all tested cases, the best *SP* is observed for *DEwPC*. It is also shown that *SR* drops as

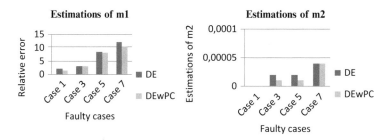

Fig. 4.8 Rage relative estimation errors for Cases 1, 3, 5, and 7

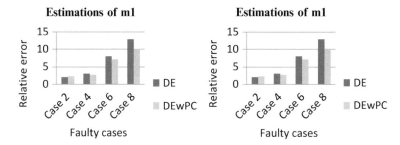

Fig. 4.9 Average relative estimation errors for Cases 2, 4, 6, and 8

Table 4.2 Results of a comparison for computational cost

		Case 1	Case 2	Case 3	Case 4	Case 5	Case 6	Case 7	Case 8
DE	SR	92	100	80	92	72	72	56	54
	SP	1913	1829	2375	2045	2756	2626	3571	3703
	SP/SPbest	1.46	1.51	1.78	1.52	1.47	1.68	1.41	1.40
DEwPC	SR	92	100	92	96	80	80	64	64
	SP	1304	1211	1336	1344	1875	1561	2534	2656
	SP/SPbest	1	1	1	1	1	1	1	1

the noise level rises. For the case of a noise level of up to 10 %, *SR* is around 64 % for *DEwPC* but less than 60 % for *DE*. This suggests that, for such a noise level, the diagnosis of the time-dependent incipient faults is not very reliable. Due to the design of the stopping criterion for these experiments, *SP* provides an indication of the number of objective function evaluations required for reaching an error of less than 2 % in the estimation of each parameter. In all cases, the *SP* from *DEwPC* is the lowest, which means that the computational cost of *DEwPC* is lower than *DE*.

Figure 4.10a–b shows the evolution of the average objective function obtained for Case 4 with *DE* and *DEwPC*, respectively. *DEwPC* reaches the lowest value for the objective function and achieves the best estimations in a number of iterations lower than *DE*.

Fig. 4.10 Evolution of the average value of the objective function, Case 4. (**a**) DE, (**b**) DEwPC

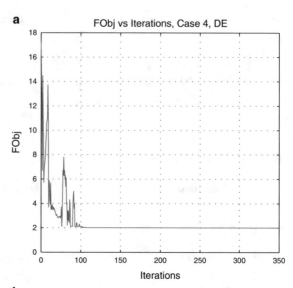

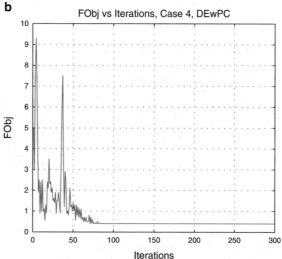

4.7   Conclusions

This chapter applies the formulation of FDI by an inverse problem methodology. The principal contribution of the chapter is that this approach is expanded to include the diagnosis of time-dependent incipient faults, as a more realistic description of practical situations.

The experiments confirmed the suitability of the inverse problem methodology; in particular, its formulation as an optimization problem, for the development of sensitive and robust FDI methods. The results have also shown that time-dependent incipient faults can be diagnosed. The comparison of the results from the two tanks system as a benchmark confirmed that *DEwPC* helps to reduce the computational cost required by *DE*.

As illustrated by the numerical experiments, the information about the noise level that affects the system can be relied upon to improve the values of the stopping criteria of the algorithms. Here is a useful tool that permits to improve the quality of fault estimations.

Acknowledgements The authors acknowledge the financial support provided by the Brazilian Agencies FAPERJ, Fundação Carlos Chagas Filho de Amparo à Pesquisa do Estado do Rio de Janeiro; CNPq, Conselho Nacional de Desenvolvimento Científico e Tecnológico; CAPES, Coordenação de Aperfeiçoamento de Pessoal de Nível Superior, as well as the Ministerio de Educación Superior de Cuba (MES).

References

1. Brest, J., Greiner, S., Boscovic, B., Mernik, M., Zumer, V.: Self-adapting control parameters in differential evolution: A comparative study on numerical benchmark problems. IEEE Trans. Evol. Comput. **10**(8), 646–657 (2006)
2. Camps-Echevarría, L., Llanes-Santiago, O., Silva Neto, A.J.: Fault diagnosis based on inverse problem solution. In: 7th International Conference on Inverse Problems in Engineering (ICIPE). Florida (2011)
3. Camps-Echevarría, L., Llanes-Santiago, O., Silva Neto, A.J.: Aplicaciónn de los algoritmos evolución diferencial y colisión de partículas al diagnóstico de fallos en sistemas industriales. Revista Investig. Oper. **18**(1), 5–17 (2012)
4. Camps-Echevarría, L., Campos Velho, H.F., Becceneri, J.C., Silva Neto, A.J., Llanes-Santiago, O.: The fault diagnosis inverse problem with ant colony optimization and fuzzy- ant colony optimization. Appl. Math. Comput. **227**(15), 687–700 (2014)
5. Camps-Echevarría, L., Llanes-Santiago, O., Silva Neto, A.J., Campos-Velho, H.F.: An approach to fault diagnosis using meta-heuristics: a new variant of the differential evolution algorithm. Comput. Sist. **18**(1), 5–17 (2014)
6. Das, S., Abraham, A., Uday, K.C., Konar, A.: Differential evolution using a neighborhood-based mutation operator. IEEE Trans. Evol. Comput. **13**(3), 526–553 (2009)
7. Ding, S.X.: Model-based Fault Diagnosis Techniques: Design Schemes, Algorithms, and Tool. Springer, London (2008)
8. Dulmage, A., Mendelsohn, N.: Coverings of bipartite graphs. Can. J. Math **10**(5), 517–534 (1958)
9. Frank, P.M.: Analytical and qualitative model-based fault diagnosis: a survey and some new results. Eur. J. Control **2**(1), 6–28 (1996)
10. Isermann, R.: Model based fault detection and diagnosis. Status and applications. Ann. Rev. Control **29**(1), 71–85 (2005)
11. Krysander, M., Frisk, E.: Sensor placement for fault diagnosis. IEEE Trans. Syst. Man Cybern. Part A Syst. Humans **38**(6), 1398–1410 (2008)
12. Krysander, M., Aslund, J., Frisk, E.: An efficient algorithm for finding minimal overconstrained subsystems for model-based diagnosis. IEEE Trans. Syst. Man Cybern. Part A Syst. Humans **38**(1), 197–206 (2008)
13. Liu, Q., Wenyuan, L.: The study of fault diagnosis based on particle swarm optimization algorithm. Comput. Inf. Sci. **2**(2), 87–91 (2009)

14. Lunze, J.: Laboratory three tanks system-benchmark for the reconfiguration problem. Technical Report, Technical University of Hamburg-Harburg, Institute of Control Engineering (1998)
15. Luz, E.F.P., Becceneri, J.C., Velho, H.F.C.: A new Multiparticle Collision Algorithm for optimization in a high-performance environment. J. Comput. Interdiscip. Sci. 1(1), 3–10 (2008)
16. Metenidin, M.F., Witczak, M., Korbicz, J.: A novel genetic programming approach to nonlinear system modelling: application to the DAMADICS benchmark problem. Eng. Appl. Artif. Intell. 17(4), 363–370 (2004)
17. Mezura-Montes, E., Velázquez, J., Coello-Coello, C.A.: A comparative study of differential evolution variants for global optimization. In: The Genetic and Evolutionary Computation Conference (GECCO). Seattle, Washington (2006)
18. Qin, A.K., Huang, V.L., Suganthan, P.N.: Differential evolution algorithm with strategy adaptation for global numerical optimization. IEEE Trans. Evol. Comput. 13(2), 398–417 (2009)
19. Sacco, W.F., Oliveira, C.R.: A new stochastic optimization algorithm based on particle collisions. In: ANS Annual Meeting. Transactions of the American Nuclear Society (2005)
20. Sacco, W.F., Oliveira, C.R., Pereira, C.M.N.: Two stochastic optimization algorithms applied to nuclear reactor core design. Prog. Nucl. Energy 48(6), 525–539 (2006)
21. Samanta, B., Nataraj, C.: Use of particle swarm optimization for machinery fault detection. Eng. Appl. Artif. Intell. 22(2), 308–316 (2009)
22. Silva Neto, A.J., Becceneri, J.C.: Técnicas de Inteligência Computacional Inspiradas na Natureza – Aplicação em Problemas Inversos e Transferência Radiativa, vol. 41, 2a Edição. Notas em Matemática Aplicada. SBMAC, São Carlos (2012) [e-ISBN 978-85-8215-002-3]
23. Simani, S., Patton, R.J.: Fault diagnosis of an industrial gas turbine prototype using a system-identification approach. Control Eng. Pract. 16(7), 769–786 (2008)
24. Simani, S., Fantuzzi, C., Patton, R.J.: Model-Based Fault Diagnosis in Dynamic Systems Using Identification Techniques. Springer, London (2002)
25. Storn, R., Price, K.: Differential evolution: A simple and efficient adaptive scheme for global optimization over continuous spaces. J. Glob. Optim. 11(4), 341–359 (1997)
26. Suganthan, P., Hansen, N., Liang, J.J., Deb, K., Chen, Y.P., Auger, A., Tiwari, S.: Problem definitions and evaluation criteria for the CEC 2005. Technical Report, Nanyang Technological University (2005); Special session on real parameter optimization
27. Torres, R.H., Mesa, M.I., Câmara, L.D.T., Silva Neto, A.J., Llanes-Santiago, O.: Application of genetic algorithms for parameter estimation in liquid chromatography. Revista Ingeniería Electrónica, Autom. Comun. 32(3), 13–20 (2011)
28. Tvrdik, J.: Adaptation in differential evolution: a numerical comparison. Appl. Soft Comput. 9(3), 1149–1155 (2009)
29. Venkatasubramanian, V., Rengaswamy, R., Yin, K., Kavuri, S.N.: A review of process fault detection and diagnosis part i: Quantitative model-based methods. Comput. Chem. Eng. 27(3), 293–311 (2003)
30. Wang, L., Niu, Q., Fei, M.: A novel quantum ant colony optimization algorithm and its application to fault diagnosis. Trans. Inst. Meas. Control. 30(3/4), 313–329 (2008)
31. Witczak, M.: Advances in model based fault diagnosis with evolutionary algorithms and neural networks. Int. J. Math. Comput. Sci. 16(1), 85–99 (2006)
32. Witczak, M.: Modeling and Estimation Strategies for Fault Diagnosis of Non-Linear Systems From Analytical to Soft Computing Approaches. Springer, London (2007)
33. Yang, E., Xian, H., Zhang, D.: A comparative study of genetic algorithm parameters for the inverse problem-based fault diagnosis of liquid rocket propulsion systems. Int. J. Autom. Comput. 4(3), 255–261 (2007)
34. Zaharie, D.: Influence of crossover on the behavior of Differential Evolution algorithms. Appl. Soft Comput. 9(3), 1126–1138 (2009)
35. Zhang, J.: Adaptive differential evolution with optional external archive. IEEE Trans. Evol. Comput. 13(5), 945–958 (2009)

Chapter 5
An Indirect Kernel Optimization Approach to Fault Detection with KPCA

José M. Bernal de Lázaro, Orestes Llanes-Santiago, Alberto Prieto-Moreno, and Diego Campos Knupp

Abstract This chapter discusses a new indirect kernel optimization criterion for the adjustment of a fault detection process that is based on the dimension–reduction technique known as kernel principal component analysis. The kernel parameter optimization proposed here involves the computation of the false alarm rate and false detection rate indicators that are combined in a single indicator: the area under the ROC curve. This approach was tested on the Tennessee Eastman (TE) process, where a significant decrease in false and missing alarms was observed.

Keywords Fault detection • Kernel PCA • Indirect optimization • AUC • Tennessee Eastman • Curve ROC historical data • False detection rate • False alarm rate

5.1   Introduction

The main objective of fault diagnosis is to detect any abnormal events and to determine their causes early enough, so that a safe and reliable operation of industrial processes will be assured [3]. With this purpose, the diagnostic process can be analyzed in three phases: detection, isolation, and identification. The detection phase compares the actual behavior of the process with its expected

J.M.B. de Lázaro (✉)
Reference Center for Advanced Education, Instituto Superior Politécnico José Antonio Echeverría (CUJAE), Marianao, La Habana, CP 19390, Cuba
e-mail: jbernal@crea.cujae.edu.cu

O. Llanes-Santiago • A. Prieto-Moreno
Automatic and Computing Department, Instituto Superior Politécnico José Antonio Echeverría (CUJAE), Marianao, La Habana, CP 19390, Cuba
e-mail: orestes@tesla.cujae.edu.cu; albprieto@electrica.cujae.edu.cu

D. Campos-Knupp
Mechanical Engineering and Energy Department, Polytechnic Institute, IPRJ-UERJ, Rua Bonfim, No. 25, Nova Friburgo, RJ, CEP 28625-570, Brazil
e-mail: diegoknupp@iprj.uerj.br

© Springer International Publishing Switzerland 2016   63
A.J. da Silva Neto et al. (eds.), *Mathematical Modeling and Computational Intelligence in Engineering Applications*, DOI 10.1007/978-3-319-38869-4_5

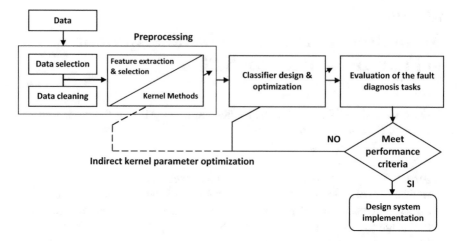

Fig. 5.1 Typical fault diagnostic schematic based on historical data with kernel methods

normal operating condition. The fault isolation and identification phases rely on the diagnostic signals (information) that are generated by detection algorithms, and as a result, the occurring faults can be identified and classified. The fault diagnosis conducted with historical data reflects a pattern recognition procedure that includes data preprocessing, as well as pattern extraction and classification, as shown in Fig. 5.1.

For complex industrial processes, data pre-processing is a critical step, since the original data are converted to a more appropriate format for the implementation of fault diagnosis tasks [14]. In addition to noise filtering and outlier removal, this step compresses the data and reduces the dimensionality of input data in a manner that the essential information contained in the data is retained, thus facilitating the fault diagnosis tasks [9]. In order to assure a high level of process monitoring performance, current researches have incorporated kernel methods in the fault detection processes [2, 4, 5, 16]. However, in these methods, parameter selection plays an essential role in designing efficient tools that can be applied to the operation of fault diagnosis tasks [5].

This chapter proposes an indirect kernel optimization approach that combines the information from false alarm rate (FAR) and false detection rate (FDR) indicators, as formulated through the area under the ROC curve (AUC). Besides to allowing one more efficient evaluation of fault detection methods, the diagnostic system's capacity is enhanced to distinguish between the normal and abnormal operating conditions. The AUC criterion is used as a kernel parameter optimization measure for adjusting the dimension–reduction technique, known as kernel principal component analysis (KPCA). The study was conducted on the test case Tennessee Eastman (TE) process, and all the techniques described herein are executed in Matlab®(R2015).

The rest of this chapter is organized as follows. In Sect. 5.2 the approach and the techniques used in the fault detection tasks are described, and the theoretical interpretation of the proposed indirect kernel optimization is discussed in detail. The optimization method that was applied to adjust the radial base function (RBF) kernel and the performance of the fault detection process in the TEP are discussed in Sect. 5.3. Finally, conclusions are drawn from result analysis, and various potential lines of research are presented.

5.2 Materials and Methods

5.2.1 Kernel Principal Component Analysis

PCA is a dimensionality-reduction technique that preserves the significant variability of the information contained in an original dataset. Its application in linear processes has shown high performance rates [9–11]. Kernel PCA is an unsupervised learning method described by Schölkopf et al. [12], where a non-linear extension of PCA is executed using the basic idea of Cover's theorem and the kernel trick [2, 5]. To derive the KPCA algorithm, let us assume that a map of a set of observations $\{\mathbf{x}_i\}_{i=1}^m \in \mathbf{R}^p$ is first generated into a possibly high dimensional dot product space $\mathcal{H}$, such that

$$\mathbf{x}_i \in \mathbf{R}^p \rightarrow \Phi(\mathbf{x}_i) \in \mathcal{H} \tag{5.1}$$

this transformation makes the mapped data more linearly separable. Then, KPCA solves the following eigenvalue problem:

$$\Sigma v = \lambda v \tag{5.2}$$

where $\Sigma = \frac{1}{m}\tilde{\Phi}(\mathbf{x}_i)\tilde{\Phi}(\mathbf{x}_j) = \frac{1}{m}K(\mathbf{x}_i, \mathbf{x}_j)$ represents the covariance matrix in $\mathcal{H}$ corresponding to m samples, assuming that $\tilde{\Phi}(\mathbf{x})$ has been previously centered. $K(\mathbf{x}_i, \mathbf{x}_j) = \langle\tilde{\Phi}(\mathbf{x}_i) \cdot \tilde{\Phi}(\mathbf{x}_j)\rangle$ is the kernel function that implicitly maps the data onto the feature space. The resulting matrix that evaluates the above function has been previously centered and scaled in $\mathcal{H}$ to obtain the orthonormal eigenvectors $\alpha_1, \alpha_2, \ldots, \alpha_d$ that correspond to the d largest positive eigenvalues $\lambda_1, \lambda_2, \ldots, \lambda_d$. Also, in order to guarantee the orthogonality of the principal components, the following condition must be satisfied:

$$\langle \alpha^j, \alpha^j \rangle = \frac{1}{\lambda_j}, \ (j = 1, \ldots, d) \tag{5.3}$$

According to [4], after the principal components have been obtained in the feature space, the jth projection of one centered value $\tilde{\Phi}(\mathbf{x}_{\text{new}})$ in $\mathcal{H}$ may be calculated as:

$$t_{\text{new},j} = \langle v_j, \tilde{\Phi}(\mathbf{x}_{\text{new}}) \rangle = \sum_{i=1}^{m} \alpha_i^j K(x_i, \mathbf{x}_{\text{new}}) \tag{5.4}$$

where $j = 1, \ldots, d$ are the principal components (PCs) in the kernel PCA. Using (5.4), a score vector $t_j = [t_{\text{new},1}, t_{\text{new},2}, \ldots, t_{\text{new},d}]^{\text{T}}$ for $\mathbf{x}_{\text{new}}$ can be obtained. It is worthwhile to point out that the SPE and Hotelling's T^2 statistics have the same interpretation as in the linear PCA. Thus, a measure of the variation within the KPCA model is given by Hotelling's T^2 statistic that is calculated using the dominant components with the maximal process variations, such that

$$T^2 = [t_1, \ldots, t_d] \Lambda^{-1} [t_1, \ldots, t_d]^{\text{T}} \tag{5.5}$$

where the score vector is obtained from (5.4) and Λ^{-1} is the diagonal matrix of the inverse of the eigenvalues associated with d as the principal components that are retained. For KPCA, the SPE statistic cannot be calculated on the residuals of the original measurement because the non-linear mapping $\tilde{\Phi}(\mathbf{x})$ is unknown. However, alternatively the residual of the mapped data can be used to calculate SPE, such that:

$$SPE = K(\mathbf{x}_{\text{new}}, \mathbf{x}_{\text{new}}) - t_{\text{new}}^{\text{T}} t_{\text{new}} \tag{5.6}$$

The threshold of Hotelling's T^2 can be calculated by the following probability distribution:

$$T^2_{\text{lim}} = \frac{q(m^2 - 1)}{m(m - q)} F_{(q, m-q, \vartheta)} \tag{5.7}$$

where $F_{q, m-q, \vartheta}$ is the F-distribution with significance level ϑ and $(m - q)$ represents the degrees of freedom, as calculated for a number of retained components q and m samples. The threshold for SPE is approximated by

$$SPE_{\text{lim}} = \theta_1 \left[\frac{h_0 c_\vartheta \sqrt{2\theta_2}}{\theta_1} + \frac{\theta_2 h_0 (h_0 - 1)}{\theta_1^2} + 1 \right]^{1/h_0} \tag{5.8}$$

$$\theta_i = \sum_{j=a+1}^{n} \lambda_j^i \qquad h_0 = 1 - \frac{2\theta_1 \theta_3}{3\theta_2^2} \tag{5.9}$$

where c_ϑ is the standard normal deviation of the $(1 - \vartheta)$ percentile and λ_j is the eigenvalue associated with the jth loading vector.

5.2.2 Indirect Kernel Optimization Criteria

The quality of a fault detection process usually entails an analysis of the number of false and missing alarms. According to [15], these indicators known as FAR and Fault Detection Rate (FDR) can be calculated as follows:

$$\text{FAR} = \frac{\text{No. of samples } (J > J_{\lim}|f = 0)}{\text{total samples } (f = 0)} \times 100 \qquad (5.10)$$

$$\text{FDR} = \frac{\text{No. of samples } (J > J_{\lim}|f \neq 0)}{\text{total samples } (f \neq 0)} \times 100 \qquad (5.11)$$

where J is any discriminative algorithm operating as binary classifier and $J_{\lim}$ is the threshold that defines whether a measurement is consistent with the normal operating condition (NOC). Despite their widespread use, the above indicators have two important disadvantages related with their supplied information. Firstly, these indicators fail to reflect the capability of the diagnostic system to distinguish between the NOC and a fault situation. In other words, each FAR and FDR represents solely the probability that NOC and abnormal operating condition (AOC) are detected, respectively. Secondly, the performance assessment of any fault detection process using these indicators requires that the values generated by FDR and FAR be analyzed simultaneously. Therefore, their interpretation can be quite subjective because the information of fault detection performance is not presented in a single numeric value.

As a way to address these disadvantages, it is hereby proposed that the interpretation of the above indicators be consolidated into a single criterion: the AUC. For this purpose, the confusion matrix C_{ij} must be used. As illustrates Fig. 5.2, each element of the confusion matrix provides a probability for the patterns of classes indicated by the row indices to be attributed to classes indicated by the column indices [7].

The matrix C_{ij} diagonal represents the number of correct samples detected/classified, and the values outside this diagonal reflect the confusion between

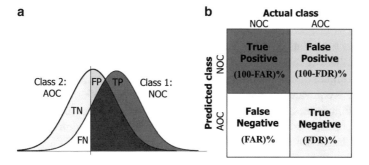

Fig. 5.2 Confusion matrix interpretation for a binary classification process. (**a**) Information obtained from the confusion matrix and separability of the PDFs corresponding with the binary classification process. (**b**) Interpretation of the confusion matrix

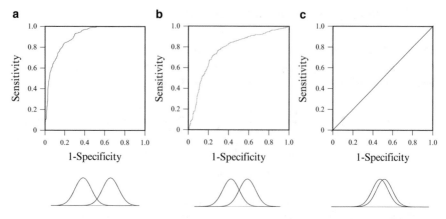

Fig. 5.3 Discriminant interpretation of the area under an ROC curve (AUC). (**a**) A high ability to distinguish between AOC and NOC. (**b**) A moderate ability to distinguish between AOC and NOC. (**c**) A poor ability to distinguish between AOC and NOC

the classes, as representative of the two operating states. Then, for any binary classification process, the following metrics can be extracted:

$$\text{Sensitivity} = \frac{TP}{TP + FN}, \quad \text{Specificity} = \frac{TN}{FP + TN} \tag{5.12}$$

Sensitivity is the proportion of positive-class samples (NOC) that are recognized as positive (TN-true positive rate expressed as a percentage). Specificity is the proportion of negative-class samples (AOC) that are recognized as negative (TN-true negative rate expressed as a percentage). The consecutively plotted values over a range of possible thresholds generate a receiver operating characteristic (ROC) curve that represents the relation between sensitivity *vs.* (1-specificity) as a two-dimensional graph. As shown in Fig. 5.3a, the AUC is larger when the difference between the classes (i.e., operating stages) is better distinguished. By increasing the confusion between classes, the AUC value decreases, as shown in Fig. 5.3b–c. Obviously, AUC $= 1$ represents a perfect detection result, and AUC ≤ 0.5 is equivalent to an FDI scheme without discriminant capacity in the detection stage.

The AUC can be calculated as the sum of the areas of the trapezoids formed by connecting the points on the ROC curve. The above represents the probability that, for a hypothetic function $g(\mathbf{x})$, a randomly selected sample $(\mathbf{x}^+)$ of the positive class will score higher than a randomly selected sample $(\mathbf{x}^-)$ of the negative class [7]. This procedure is equivalent to applying the Wilcoxon–Mann–Whitney test in the binary classification process. Then, in a fault detection process that involves c mutually exclusive operating states other than NOC, the AUC evaluation criterion can be used as a fitness function, as follows:

$$\textit{Fitness function} = 1 - \frac{1}{c} \sum_{i=1}^{c} \text{AUC}^i \tag{5.13}$$

The basic concept underpinning this criterion is to minimize the probability of error for the discrimination process of each pair (NOC, AOC^i), with $i = \{1, 2, \ldots, c\}$. For instance, as illustrated in Fig. 5.3a–b, let us assume that *Class* 2 represents three mutually exclusive abnormal operating states associated with faults $\{f_1, f_2, f_3\}$. An AUC value is obtained for each binary detection process, such that $AUC^i = \{0.85, 0.75, 0.5\}$, respectively. Then, *Fitness function* $= 1 - \frac{1}{3}\{0.85 + 0.75 + 0.5\} = 0.3$ can be interpreted as the probability rate that NOC be confused with AOCs. In order to achieve high fault detection performance, AUC ≈ 1 must be satisfied for each binary classification process. In other words, maximizing the AUC measure of the detection process for any individual faults is equivalent to minimizing the overall misclassification rate in the fault detection stage.

5.3 Optimization with Differential Evolution

Differential evolution (DE) is one of the most popular optimization algorithms, thanks to its good convergence and fast implementation. Originally proposed by Storm and Price [13], DE has proven to be a very reliable optimization strategy for many different tasks [1, 8]. This algorithm is based on three operators; i.e., Mutation, Crossover, and Selection. These operators require that a population size (NP), the number of parameters to be optimized, and a scale factor (F) are defined. The crucial idea behind DE is to combine these operators at each *j*th iteration, using vector operations, in order to obtain a new solution candidate. The following equation shows the manner in which this algorithm combines three population members x_{ri}, $i = \{0, 1, 2\}$, to create a new solution $v_{i,g}$ from the current generation g:

$$v_{j,i,g} = x_{j,r0,g} + F \times (x_{j,r1,g} - x_{j,r2,g}) \tag{5.14}$$

where $F \in (0, 1)$ is the scale factor that determines the rate at which the population evolves, x_{r0} is the initial candidate position and $\{x_{r1}, x_{r2}\}$ is the search range to the *j*th iteration, respectively. The pseudo-code in Fig. 5.4 provides a general description of the DE algorithm.

Input: *NP*, *MaxIter*, *F*, crossover criteria, selection criteria and search range;
Output: Position of the approximate global optima $\mathbf{X}^{best}$
1: **for** *Iter* $\leftarrow 1, MaxIter$ **do**
2:     Evaluate the parameters;
3:     Apply *Mutation*, *Crossover* and *Selection*
4:     Evaluate fitness:= $f(x_{ri})$;
5:     Update X^{best}
6:     Verify stopping criteria
7: **end for**
8: Solution: $\mathbf{X}^{best}$

Fig. 5.4 Pseudo-code of the algorithm differential evolution

The operating mechanism of this algorithm can be summarized with the following notation:

$$DE/X^{\delta}_{\text{Iter}-1}/\gamma/\lambda \qquad (5.15)$$

where γ indicates the number of pairs of solutions from the current population to be used for perturbing the present solution $X^{\delta}_{\text{Iter}-1}$; λ represents a distribution function to be used during *Crossover*.

In this chapter, the $DE/X^{\text{best}}/2$ bin scheme was employed, where *bin* is a notation for a binomial distribution function used during the crossover. The implementation of this algorithm is available at http://www.icsi.berkeley.edu/~storn/code.html.

5.4   Results and Discussion

In this section, the proposed indirect kernel optimization criterion and the above-mentioned techniques are applied in the Tennessee Eastman (TE) process for the purpose of detecting the simulated faults in this industrial benchmark.

5.4.1   Study Case: Tennessee Eastman Process

The Tennessee Eastman (TE) process is widely used as a chemical benchmark to assess the performance of new monitoring and control strategies [5, 9, 15]. The process consists of five major units; i.e., a reactor, a condenser, a recycle compressor, a separator, and a stripper. All these units are interconnected, as shown in the schematic flow diagram in Fig. 5.5.

The TE process contains 21 pre-programmed faults and one NOC dataset. Historical datasets are generated for 48 h, and faults are introduced after 8 h of simulation have elapsed. Each historical dataset contains a total of 52 variables, with a 3-min sampling time and Gaussian noise incorporated in every measurement. The control objectives and general features of this process simulation are described in detail by Chiang et al. [3] and Downs and Vogel [6]. For the purpose of assessing the benefits of the proposal described herein, only the 33 online available variables of the above-mentioned process were considered. The designations of these variables are listed in Table 5.1. The datasets used to test the proposed procedures were supplied by Downs and Vogel [6] and may be downloaded from http://web.mit.edu/braatzgroup/TE_process.zip.

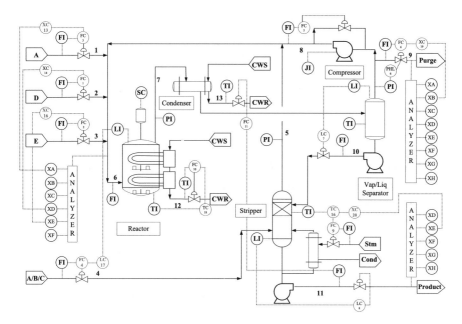

Fig. 5.5 Instrumentation diagram of the Tennessee Eastman process [3]

Table 5.1 Description of faults in TE process

Fault	Process variable	Type	Fault	Process variable	Type
F1	A/C feed ratio, B composition constant	Step	F9	D feed temperature	Random
F2	B composition, A/C ration constant	Step	F10	C feed temperature	Random
F3	D feed temperature	Step	F11	Reactor cooling water inlet	Random
F4	Reactor cooling water inlet temperature	Step	F12	Temperature	Random
F5	Condenser cooling water inlet	Step	F13	Reaction kinetics	Slow drift
F6	A feed loss	Step	F14	Reactor cooling water valve	Sticking
F7	C header pressure loss-reduced availability	Step	F15	Condenser cooling water valve	Sticking
F8	A, B, and C feed composition	Random			

5.4.2 Experimental Description

For the purpose of monitoring the faults of the TE process, the KPCA and the SPE and Hotelling's T^2 statistics described in the preceding sections were applied. The differential evolution (DE) algorithm was employed for the kernel parameter optimization with the following specifications: population size $NP = 10$, maximum iteration $MaxIter = 200$, difference vector scale factor $F = 0.75$, and crossover criterion $CR = 0.6$.

From the fault detection process, the AUC was used as the fitness function in the optimization algorithm DE (See (5.13)). The stopping criteria that were applied herein included the number of iterations and the value reached by the objective function. The search range for the parameter of the RBF kernel was $\sigma \in [300, 1500]$, and its resulting value was $\sigma = 340.40553$. Based on this, the number of principal components utilized in KPCA was 19 PCs, which explains 95.07% of the total process information. Consequently, the feature space dimension was reduced from $\mathbf{R}^{33}$ to $\mathbf{R}^{19}$, according to the 0.001 cutoff value for the eigenvalues.

Since the effect of a fault does not have to be simultaneously reflected in systematic and noisy parts of the KPCA model, the information of the SPE and Hotelling's T^2 statistics was combined as shown below:

$$\left[(T^2(x_i) > T^2_{\text{lim}}) \lor (\text{SPE}(x_i) > \text{SPE}_{\text{lim}})\right] \iff (f \neq 0) \qquad (5.16)$$

As illustrated by Fig. 5.6, if the value of either statistics exceeds its threshold, then a fault is deemed to have occurred. For both SPE and Hotelling's T^2 statistics, a 99% confidence limit has been adopted as the alarming threshold.

5.4.3 Discussion

Once the configuration of the fault detection system has been defined, an online detection process may be conducted using the testing dataset of the TE process. As illustrated by Fig. 5.7 the results obtained from the application of KPCA using the aforementioned specifications are expressed in terms of the AUC measure. The analysis of the results has shown that certain faults in the TE process can be easily detected (e.g., Faults 1, 2, 4, 6, 7, 8, 12, and 14) for their magnitude that departs from the NOC of the process. Nonetheless, other faults are harder to detect (e.g., Faults 3, 5, 9, 10, 11, 13, and 15).

By applying the proposal contained herein, the fault detection performance rate exceeds 90% for Faults 1, 2, 4, 6, 7, 8, 12, and 14. However, the detection performance rate for Faults 3, 5, 9, 10, 11, 13, and 15 is lower than 70% because not major changes are observed in the mean, the variance and the higher-order statistics,

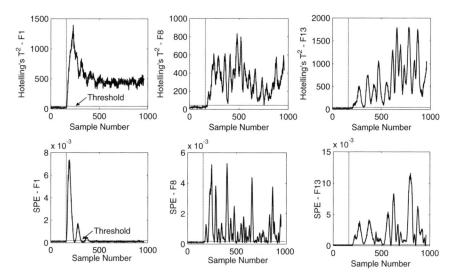

Fig. 5.6 Process monitoring charts for the TE process in the case of Faults 1, 8, and 13

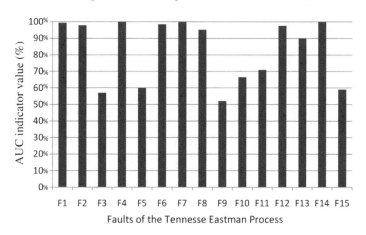

Fig. 5.7 AUC performance obtained from the online fault diagnosis process

which may have otherwise be relied upon to tell their differences from the NOC. Due to their small order of magnitude, these poorly detectable faults may be concealed under the influence of other variables in the process. The application of conventional detection approaches may result in poor performance for such faults. In other words, the detection of Faults 3, 9, and 15 requires that the dynamic behavior of the process involved is taken into account.

5.5 Conclusions

This chapter presents an assessment criterion that combines FAR and FDR indicators formulated as the AUC. This criterion was introduced as kernel parameter optimization measure to adjust the parameters of the dimension–reduction technique KPCA using differential evolution (DE). As noted above, the FAR and the Fault Detection Rate (FDR) are often used to appraise the quality of fault detection techniques. However, the FAR and FDR indicators describe solely the percentage of samples detected on the basis of NOC and AOC datasets, but their individual interpretation does not reflect the diagnostic system's capability to distinguish between these operating states. The use of the AUC measure to test fault detection results is a more appropriate option. A high AUC value represents a strong discriminative capacity, and hence, a high probability that a normal-operation sample may have been told apart from an observed fault situation. Experiments have shown that the combination of the proposed approach with this kernel method helps to attain high performance rates in the execution of fault detection tasks.

5.6 Future Work

Future research may focus on the application of other detection algorithms that consider process dynamics for the purpose of improving the detection of incipient and small magnitude faults. It would also be interesting to investigate the use of other optimization techniques that help in the adjustment of the kernel parameters.

Acknowledgements The authors thank the financial support provided by the Brazilians Agencies FAPERJ, Fundação Carlos Chagas Filho de Amparo à Pesquisa do Estado do Rio de Janeiro; CNPq, Conselho Nacional de Desenvolvimento Científico e Tecnológico; CAPES, Coordenação de Aperfeiçoamento de Pessoal de Nível Superior, and MES/CUBA, Ministerio de Educación Superior de Cuba.

References

1. Camps-Echevarría, L., Llanes-Santiago, O., Silva Neto, A.J., Campos-Velho, H.F.: An approach to fault diagnosis using meta-heuristics: a new variant of the differential evolution algorithm. Comput. Sist. **18**(1), 5–17 (2014)
2. Cheng, C., Hsu, C., Chen, M.: Adaptive kernel principal component analysis (KPCA) for monitoring small disturbances of nonlinear processes. Ind. Eng. Chem. Res. **49**(5), 2254–2262 (2010)
3. Chiang, L., Braatz, R., Russell, E.: Fault detection and diagnosis in industrial systems. Springer, London (2001)
4. Choi, S., Lee, C., Lee, J., Park, J., Lee, I.: Fault detection and identification of nonlinear processes based on kernel PCA. Chemom. Intell. Lab. Syst. **75**(1), 55–67 (2005)

5. de Lázaro, J.B., Prieto-Moreno, A., Llanes-Santiago, O., Silva-Neto, A.J.: Optimizing kernel methods to reduce dimensionality in fault diagnosis of industrial systems. Comput. Ind. Eng. **87**, 140–149 (2015)
6. Downs, J., Vogel, E.: A plant-wide industrial process control problem. Chem. Eng. **17**(3), 245–255 (1993)
7. Fawcett, T.: An introduction to ROC analysis. Pattern Recogn. Lett. **27**(8), 861–874 (2006)
8. Knupp, D., Sacco, W., Silva-Neto, A.: Direct and inverse analysis of diffusive logistic population evolution with time delay and impulsive culling via integral transforms and hybrid optimization. Appl. Math. Comput. **250**, 105–120 (2015)
9. Prieto-Moreno, A., Llanes-Santiago, O., de Lázaro, J.B., Garcia-Moreno, E.: Comparative evaluation of classification methods used in fault diagnosis of industrial processes. IEEE Lat. Am. Trans. **11**(2), 682–689 (2013)
10. Prieto-Moreno, A., Llanes-Santiago, O., García-Moreno, E.: Principal components selection for dimensionality reduction using discriminant information applied to fault diagnosis. J. Process Control **33**, 14–24 (2015)
11. Qin, S.: Survey on data-driven industrial process monitoring and diagnosis. Ann. Rev. Control **36**(2), 220–234 (2012)
12. Schölkopf, B., Smola, A., Müller, K.: Nonlinear component analysis as a kernel eigenvalue problem. Neural Comput. **10**(5), 1299–1319 (1998)
13. Storn, R., Price, K.: Differential evolution-a simple and efficient adaptive scheme for global optimization over continuous spaces. J. Glob. Optim. **11**(4), 341–359 (1997)
14. Venkatasubramanian, V., Rengaswamy, R., Yin, K., Kavuri, S.: A review of process fault detection and diagnosis : Part I: Quantitative model based methods. Comput. Chem. Eng. **27**(3), 293–311 (2003)
15. Yin, S., Ding, S., Haghani, A., Hao, H., Zhang, P.: A comparison study of basic data-driven fault diagnosis and process monitoring methods on the benchmark Tennessee Eastman process. J. Process Control **22**(9), 1567–1581 (2012)
16. Zhang, Y.: Enhanced statistical analysis of nonlinear processes using KPCA, KICA and SVM. Chem. Eng. Sci. **64**(5), 801–811 (2009)

Chapter 6
Uncertainty Quantification in Chromatography Process Identification Based on Markov Chain Monte Carlo

Mirtha Irizar Mesa, Leôncio D. Tavares Câmara, Diego Campos-Knupp, and Antônio José da Silva Neto

Abstract Modeling and simulation of chromatography systems leads to better understanding of the mass transfer mechanisms and operational conditions that can be used to improve molecular separation/purification. In this chapter, parameter uncertainty produced by the model and measurement errors in a front velocity chromatography model is quantified by means of a Bayesian method, the delayed rejection adaptive metropolis algorithm, which is a variant of the Markov Chain Monte Carlo (MCMC) method. The model is also evaluated for a random sample of parameters, being then determined the uncertainty in the prediction.

Keywords Chromatography • Parameter estimation • Uncertainty • Bayesian techniques • Markov chain Monte Carlo • Delayed rejection Adaptive metropolis algorithm • Convergence assessing methods

6.1 Introduction

Modeling and simulation of chromatography systems leads to a better understanding of the mass transfer mechanisms and operational conditions that can be used to improve molecular separation/purification. Inverse problem techniques application in the simulated moving bed (SMB) process optimization is a new trend in modeling and simulation [10–12], which consists on a complex process for the determination of more adequate conditions to obtain higher products purity. Modeling of column chromatography allows the determination of the performance of final separation

M.I. Mesa (✉)
Automatic and Computing Department, Instituto Superior Politécnico José Antonio Echeverría (CUJAE), Marianao, La Habana, CP 19390, Cuba
e-mail: mirtha@electrica.cujae.edu.cu

L.D. Tavares Câmara • D. Campos-Knupp • A.J. Silva Neto
Mechanical Engineering and Energy Department, Polytechnic Institute, IPRJ-UERJ,
Rua Bonfim, No. 25, Nova Friburgo, RJ, CEP 28625-570, Brazil
e-mail: dcamara@iprj.uerj.br; diegoknupp@iprj.uerj.br; ajsneto@iprj.uerj.br

© Springer International Publishing Switzerland 2016
A.J. da Silva Neto et al. (eds.), *Mathematical Modeling and Computational Intelligence in Engineering Applications*, DOI 10.1007/978-3-319-38869-4_6

of SMB, because such process corresponds to various interconnected columns. Different research groups have used dispersion models [5, 6] to represent chromatographic columns. Those models are robust and efficient, but need a numerical treatment of their partial differential equations, which requires a high computational time. In this chapter, columns that are utilized in an SMB system are modeled through a new approach. It consists in the use of the experimental volumetric flux to discretize the volumetric elements which move along the porous column, being the molecular transport in the liquid phase the main phenomenon in the chromatographic column, followed by the mass transfer between solid adsorbent and liquid phase. In the present chapter the parameter uncertainty produced by the model and measurement errors in a front velocity chromatography model is quantified by means of a Bayesian method, the delayed rejection adaptive metropolis (DRAM) algorithm, which is a variant of the Markov Chain Monte Carlo (MCMC) method. The model is also evaluated for a random sample of parameters, being determined the uncertainty in the prediction. The chapter summarizes the new model characteristics and the MCMC method, specifically the DRAM algorithm [8], for uncertainty quantification in the estimation of global adsorption and desorption kinetics constants of mass transfer. Synthetic data are utilized and satisfactory results are shown.

6.2 Front Velocity Chromatography Model

In this chapter a new chromatography model is applied, the convective front velocity proposed by Câmara [1], that uses the experimental velocity of the liquid phase for the discretization of volumetric elements that move along the porous column. In this approach the liquid phase convection is considered the main phenomenon in molecules transport in chromatographic columns, followed by the mass transfer between the solid adsorbent and the liquid phase. Liquid phase transmission is considered as the main phenomenon due to a flux velocity in the column which is driven by an external pump system. The required time for the liquid phase to percolate the column can be determined because the volumetric flux rate, porosity, and the column volume are experimentally known. In the chromatographic column shown in Fig. 6.1, the control volume of size J^* moves along the column with the same velocity of the effluent flux. The column length is discretized with control volumes of size J^*. The time interval (Δt) for the liquid phase movement along each control volume is obtained in (6.1)

$$\Delta t = \frac{\varepsilon V}{nF} \tag{6.1}$$

In this equation ε, V, n and F correspond to the column bed porosity, total volume of the column, total number of control volumes, and the liquid flow rate, respectively.

Fig. 6.1 Chromatographic column of length L with a discretization volume of size J^*

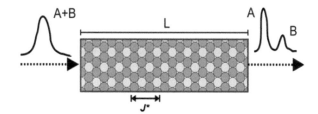

In the step of kinetic mass transfer two models are assumed. In the first one (6.2) there is not a maximum capacity of adsorption (q_m), while in the second model (6.3) this capacity is incorporated by means of Langmuir kinetics. In these models C, q, q_m, k_1 and k_2 are, respectively, the concentration in the liquid and solid phases, the maximum capacity of adsorption, and the global constants of adsorption and desorption.

$$\frac{dq}{dt} = k_1 C - k_2 q \tag{6.2}$$

$$\frac{dq}{dt} = k_1 C (q_m - q) - k_2 q \tag{6.3}$$

The kinetic models previously described drive, in equilibrium, to linear and Langmuir isotherms, respectively.

In order to estimate unknown parameters in the previously described adsorption model, the inverse problem is formulated as an optimization one, with an objective function obtained by the sum of squared residuals between experimental and calculated values of solute concentration C.

Model parameters k_1 and k_2 (6.3) are determined in this chapter by means of a Bayesian method [4], the MCMC algorithm, that yields posterior probability distribution functions for each unknown parameter instead of a punctual estimation. A variant of such algorithm, the DRAM was used.

6.3 Parameter Uncertainty Quantification

Bayes rule relates a posterior probability density distribution $\pi \left(\dfrac{\vec{\theta}}{\vec{C_e}} \right)$ to a previous probability density distribution $\pi_{\text{prev}} \left(\vec{\theta} \right)$ according to (6.4),

$$\pi \left(\frac{\vec{\theta}}{\vec{C_e}} \right) = \frac{p \left(\dfrac{\vec{C_e}}{\vec{\theta}} \right) \pi_{\text{prev}} \left(\vec{\theta} \right)}{\int_{\mathbf{R}^d} p \left(\dfrac{\vec{C_e}}{\vec{\theta}} \right) \pi_{\text{prev}} \left(\vec{\theta} \right) d\theta} \tag{6.4}$$

where $\vec{\theta} \in \mathbf{R}^d$ are the model parameters (k_1 and k_2), and $p\left(\frac{\vec{C_e}}{\vec{\theta}}\right)$ represents the observation possibility of experimental data $\vec{C_e}$ from parameters $\vec{\theta}$ [4],

$$p\left(\frac{\vec{C_e}}{\vec{\theta}}\right) = exp\left(-0.5\sigma^2 SC\left(\vec{\theta}\right)\right) \tag{6.5}$$

being σ the measurement error standard deviation, $SC\left(\vec{\theta}\right)$ is calculated as:

$$SC\left(\vec{\theta}\right) = \sum_{i=1}^{N}\left[C_e(t_{(i)}) - C_m\left(t_{(i)}; \vec{\theta}\right)\right]^2$$

and represents the sum of quadratic error between experimental observations $C_e(t_{(i)})$ and model output $C_m\left(t_{(i)}; \vec{\theta}\right)$, and N is the total number of experimental data.

6.3.1 Delayed Rejection Adaptive Metropolis Algorithm

For large parameter spaces of dimension d, the calculation of the integral in (6.4) is computationally expensive. MCMC methods avoid the difficulty to integrate the probability directly by creating a Markovian chain whose stationary distribution is the posterior density. Posterior probability density is the conditional probability density of the parameters given by the measurements. In many cases the prior parameters distribution do not yield an analytical procedure for the posterior distribution. MCMC methods are then used to sample the possible parameters and infer the posterior density from the samples [13]. Usually the number of samples required to approximate a probability distribution and the computational cost can be large. MCMC is an iterative version of Monte Carlo methods. It is based on a posterior distribution sample and the calculation of estimative samples from this distribution using iterative simulation. A Markov chain is a stochastic process $\{\theta_0, \theta_1, \ldots\}$ in which θ_i distribution, given all previous values $\theta_0, \ldots, \theta_{i-1}$, depends only of θ_{i-1}. Mathematically,

$$P\left(\theta_i \in \mathbf{A} \mid \theta_0, \ldots, \theta_{i-1}\right) = P\left(\theta_i \in \mathbf{A} \mid \theta_{i-1}\right) \tag{6.6}$$

for any subset $\mathbf{A}$. The transition probability $P(i,j) = P(i \rightarrow j)$ defines the Markov chain and represents the probability that a process moves from (state i) S_i to S_j (state j) in only one step.

To obtain a unique equilibrium distribution, MCMC methods require the following chain characteristics:

Homogeneity: Transition probabilities from one to another state are invariants.
Irreducibility: Each state can be reached from any other in a finite number of
iterations.
Aperiodicity: To avoid generating a Markov chain that remains in an infinite
periodic loop indefinitely.

So, a sufficient condition to get a unique stationary distribution is that the process
attends the balance equation

$$P(i \rightarrow j)p_i(\theta \mid X) = P(j \rightarrow i)p_j(\theta \mid X) \tag{6.7}$$

$p_i(\theta \mid X)$ and $p_j(\theta \mid X)$ represent different states.

While iterations number grows the chain forgets the initial values and eventually
converges to an equilibrium distribution. So, it is common to discard initial iterations
in practical applications, which are considered to be in the burn in stage. Some of
the most used algorithms to generate Markov chains whose distributions converge
to the distributions of interest are Metropolis Hastings and Gibbs Sampler.

In Metropolis Hastings algorithms a value is generated using an auxiliary distribution from which samples can be easily obtained and accepted with an established
probability. This correction mechanism guarantees the chain convergence to an
equilibrium distribution.

There are many MCMC algorithms, including the popular Metropolis–Hastings
algorithm [2, 14, 15] and adaptive algorithms [7, 8]. In this chapter the DRAM
algorithm presented by Haario et al. [8] is used, whose implementation is available
in [9]. MCMC algorithms create a Markovian Chain, or random walk through the
parameter space. Metropolis algorithm proposes new parameters $\vec{\theta}^{\,new}$ based on
actual parameters $\vec{\theta}^{\,ac}$ and a proposed function $f\left(\dfrac{\vec{\theta}^{\,new}}{\vec{\theta}^{\,ac}}\right)$. Standard Metropolis
algorithm [14] uses as a proposal that $\vec{\theta}^{\,new} = \vec{\theta}^{\,ac} + Az^{-1}$, being $\vec{z}$ a random
vector sampled from standard normal distribution, and A is the Cholesky factor of
covariance matrix. The adaptive Metropolis algorithm updates the covariance based
on the past trajectory of parameters chains. Adaptation ensures a good mixture of
chains or an efficient parameter space exploration. After the proposal of a new
parameter set, $\vec{\theta}^{\,new}$, parameters may be accepted with the probability calculated
in (6.8),

$$\alpha = min\left(1, \frac{p\left(\dfrac{\vec{C}_e}{\vec{\theta}^{\,new}}\right) f\left(\dfrac{\vec{\theta}^{\,ac}}{\vec{\theta}^{\,new}}\right) \pi_{prev}\left(\vec{\theta}^{\,new}\right)}{p\left(\dfrac{\vec{C}_e}{\vec{\theta}^{\,ac}}\right) f\left(\dfrac{\vec{\theta}^{\,new}}{\vec{\theta}^{\,ac}}\right) \pi_{prev}\left(\vec{\theta}^{\,ac}\right)}\right) \tag{6.8}$$

In the standard Metropolis algorithm, if $\vec{\theta}^{\,new}$ is rejected, the new proposal is
based on $\vec{\theta}^{\,ac}$ only. With the delayed rejection, the new proposal is based in $\vec{\theta}^{\,ac}$ and
the rejected parameters. The process can be repeated until a new set of parameters

Table 6.1 Parameter values
for the model direct solution
in (6.3)

k_1 subst 1	0.01798 L/(g.min)
k_2 subst 1	0.03001 min^{-1}
k_1 subst 2	0.01302 L/(g.min)
k_2 subst 2	0.01098 min^{-1}

is accepted or until an iteration number limit is reached, in which the proposal is based on $\vec{\theta}^{\text{ac}}$ only. DRAM algorithm combines delayed rejection and adaptation, incorporating learned information on the posterior distribution during the Markov chain progress, which constitutes the adaptation procedure, and realizes a second stage of searching if a candidate is rejected.

DRAM algorithm includes the following steps:

1. Start from an initial value θ_0 and initial first stage proposal covariance $C_{(i)} = C_0$. Select the scaling factor s, covariance regularization factor κ, initial non-adaptation period n_0, and scalings for the higher-stage proposal covariances $C_{(i)}, i = 1, \ldots, Ntry$, where $Ntry$ is the number of tries allowed.
2. *DR* loop. Until a new value is accepted, or $Ntry$ tries have been made:

    a. Propose θ_* from a Gaussian distribution centered at the current value $N\theta_{(i-1)}, C_{(k)}$.
    b. Accept according to the $k'th$ stage acceptance probability.

3. Set $\theta_i = \theta_*$ or $\theta_i = \theta_{i-1}$, according to whether the value is accepted or not.
4. After an initial period of simulation $i \geq n_0$, adapt the master proposal covariance using the chain generated so far $C_{(1)} = cov\,(\theta_0 \ldots \theta_i)\,s + I\kappa$. Calculate the higher-stage proposal as scaled versions of $C_{(1)}$, according to the chosen rule.
5. Iterate from 2 onwards until enough values have been generated.

    To estimate k_1 and k_2 parameters applying *DRAM* synthetic data are used, which are obtained evaluating directly the front velocity chromatography model for parameter values shown in Table 6.1 and considering two substances (subst 1 and subst 2) which could be, for example, glucose and fructose.

In Fig. 6.2 the MCMC samples for the parameters k_1 and k_2 are shown, in correspondence with the number of algorithm iterations. A good agreement with the parameter values in Table 6.1 is observed. The initial discarded or burn in samples correspond to 500 iterations.

6.4 Convergence Diagnostics

Although from theory Markov chains are expected to eventually converge to a stationary distribution, which is the target distribution, there is no guarantee that the chain has converged after M draws. So, several tests can be done, both visual and statistical, to see if the chain has indeed converged [3].

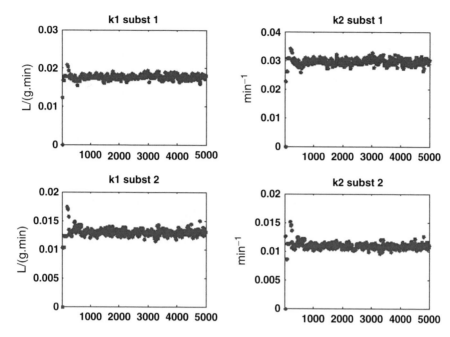

Fig. 6.2 Markov chain for the parameters k_1 and k_2

One way to analyze if the chain has converged to see how well it is mixing, or moving around the parameter space. If the chain is taking a long time to move around the parameter space, then it will take longer to converge. It is possible to see how well the chain is mixing through visual inspection for every parameter. When plotting the iteration number against the value of the draw of the parameter at each iteration the chain gets stuck in certain areas of the parameter space, it indicates bad mixing.

Running mean plots to check how well the chains are mixing can also be used. A running mean plot is a plot of the iterations against the mean of the draws up to each iteration.

Another way to assess convergence is to assess the autocorrelations between the draws of Markov chain. If autocorrelation is still relatively high for higher values of lags, this indicates high degree of correlation between draws and slow mixing.

Figure 6.3 shows the chains autocorrelation functions for the four estimated parameters.

Among others, Gelman and Rubin multiple sequence diagnostic, Geweke diagnostic, Raftery and Lewis diagnostic, and Heidelberg and Welch diagnostic are statistical methods to asses chain convergence [3].

The Heidelberg and Welch diagnostic is based on the Cramer–von Mises statistical test to accept or reject the null hypothesis that the Markov chain is from a stationary distribution. This diagnostic consists of two parts.

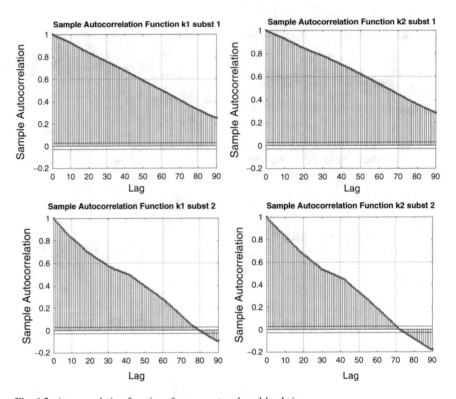

Fig. 6.3 Autocorrelation functions for parameters k_1 and k_2 chains

First Part

1. Generate a chain of N iterations and define a β level.
2. Calculate the statistical test on the whole chain. Accept or reject null hypothesis that the chain is from a stationary distribution.
3. If null hypothesis is rejected, discard the first 10 % of the chain. Calculate the statistical test and accept or reject null.
4. If null hypothesis is rejected, discard the next 10 % and calculate the statistical test.
5. Repeat until null hypothesis is accepted or 50 % of the chain is discarded. If test still rejects null hypothesis, then the chain fails the test and needs to be run longer.

Second Part

If the chain passes the first part of the diagnosis, it is used the part of the chain not discarded from the first part to test the second part. The halfwidth test calculates half the width of the $(1 - \beta)$ credible interval around the mean. If the ratio of the halfwidth and the mean is lower than some value ζ, then the chain passes the test. Otherwise, the chain must be run out longer.

Table 6.2 Heidelberg and Welch Diagnostic p-value results

k_1 subst 1	k_2 subst 1	k_1 subst 2	k_2 subst 2
0.1856×10^{-5}	0.1434×10^{-5}	0.2562×10^{-5}	0.3460×10^{-5}

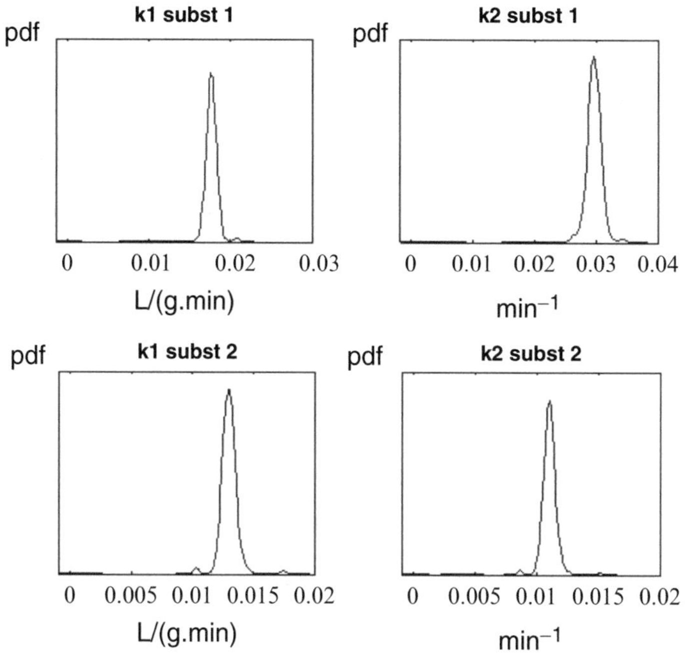

Fig. 6.4 Probability distribution functions for the estimated parameters k_1 and k_2

This last method was applied to determine the chains convergence for parameters k_1 and k_2. According to p-values shown in Table 6.2 and autocorrelation functions of Fig. 6.3, it is possible to conclude that the chains have converged.

Figure 6.4 corresponds to the posterior densities for k_1 and k_2 parameters of substance 1 and k_1 and k_2 of substance 2.

The higher probability region is located in a little fraction of parameter space, corresponding to the real values of parameters to estimate.

In Fig. 6.5 experimental data [1],and the calculated concentration values obtained by the model are compared, showing a satisfactory agreement of model evaluation for the mean values of estimated parameters k_1 and k_2.

Figure 6.6 was obtained assessing the model for a random sample of parameters, and represents prediction uncertainty. Predictive plots are obtained evaluating the model for 500 random samples from the chain.

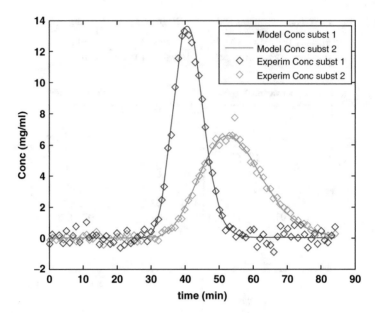

Fig. 6.5 Comparison between experimental data and those obtained by the model

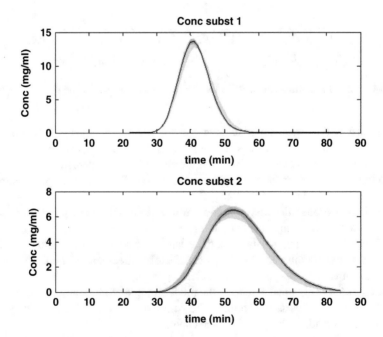

Fig. 6.6 Uncertainty prediction

6.5 Conclusions

MCMC methods are tools that yield to the estimation of parameter densities instead of their punctual estimation. In the front velocity chromatography model the estimation performed led to satisfactory results for the adsorption and desorption kinetic constants of mass transfer. Uncertainty in model prediction was also presented, showing the usefulness of this method. The aspects that drive to estimation errors are themes for future researching.

Acknowledgements The authors acknowledge the financial support provided by the Brazilian Agencies FAPERJ, Fundação Carlos Chagas Filho de Amparo à Pesquisa do Estado do Rio de Janeiro; CNPq, Conselho Nacional de Desenvolvimento Científico e Tecnológico and CAPES, Coordenação de Aperfeiçoamento de Pessoal de Nível Superior, as well as the Ministerio de Educación Superior de Cuba (MES/Cuba).

References

1. Câmara, L.D.T.: Chromatographic columns characterization for SMB (Simulated Moving Bed) separation of glucose and fructose. In: 8th European Congress of Chemical Engineering- ECC (2011)
2. Chib, S., Greenberg, E.: Understanding the Metropolis-Hastings algorithm. Am. Stat. **49**(4), 327–335 (1995)
3. Cowles, M.K., Carlin, B.P.: Markov chain Monte Carlo convergence diagnostics: a comparative review. J. Am. Stat. Assoc. **91**(434), 883–904 (1996)
4. Gelman, A., Carlin, J., Stern, H., Rubin, D.: Bayesian Data Analysis, 2nd edn. Chapman and Hall, London (2004)
5. Gu, T., Tsai, G.J., Tsao, G.T.: Modeling of nonlinear multicomponent chromatography. In: Tsao, G.T. (ed.) Advances in Biochemical Engineering/Biotechnology, vol. 49, pp. 45–71. Springer, Berlin (1993)
6. Guiochon, G.: Preparative liquid chromatography. J. Chromatogr. A. **965**(1–2), 129–161 (2002)
7. Haario, H., Saksman, E., Tamminen, J.: An adaptive metropolis algorithm. Bernoulli **7**(2), 223–242 (2001)
8. Haario, H., Laine, M., Mira, A., Saksman, E.: DRAM: efficient adaptive MCMC. Stat. Comput. **16**(4), 339–354 (2006)
9. Laine, M.: MCMC Toolbox for Matlab. MathWorks (2007). http://helios.fmi.fi/~lainema/mcmc/
10. Lee, K.B., Kasat, R., Cox, G., Wang, N.: Simulated moving bed multiobjective optimization using standing wave design and genetic algorithm. Am. Inst. Chem. Eng. J. **54**(11), 2852–2871 (2008)
11. Nam, H.G., Mun, S.: Optimal design and experimental validation of a three-zone simulated moving bed process based on the Amberchrom-CG161C adsorbent for continuous removal of acetic acid from biomass hydrolyzate. Process Biochem. **47**, 725–734 (2012)
12. Nam, H.G., Park, C., Jo, S.H., Suh, Y.W., Mun, S.: Continuous separation of succinic acid and lactic acid by using a three-zone simulated moving bed process packed with Amberchrom-CG300C. Process Biochem. **47**, 2418–2426 (2012)

13. Orlande, H.R.B., Colaço, M.J., Naveira-Cotta, M.J., Guimãraes, C., Borges, V.L. (eds.): Método de Monte Carlo com Cadeia de Markov. Inverse Problems in Heat Transfer (in Portuguese), Notas em Matemática Aplicada, vol. 51, pp. 67–80. Sociedade Brasileira de Matemática Aplicada e Computacional (www.sbmac.org.br) - SBMAC (2011)
14. Roberts, G.O., Rosenthal, J.S.: Optimal scaling for various Metropolis-Hastings algorithms. Stat. Sci. **16**(4), 351–367 (2001)
15. Solonen, A.: Monte Carlo Methods in Parameter Estimation of Nonlinear Models. Master's Thesis, Lappeenranta University of Technology (2006)

Chapter 7
Inverse Analysis of a New Anomalous Diffusion Model Employing Maximum Likelihood and Bayesian Estimation

Diego Campos-Knupp, Luciano G. da Silva, Luiz Bevilacqua, Augusto C.N.R. Galeão, and Antônio José da Silva Neto

Abstract The classical diffusion equation models the behavior of several physical phenomena related to dispersion processes quite successfully; however, in some cases, this approach fails to represent the actual physical behavior. In most published works dealing with this situation, the well-known second order parabolic equation is assumed as the basic governing equation of the dispersion process, but the anomalous diffusion effect is modeled with the introduction of fractional derivatives or the imposition of a convenient variation of the diffusion coefficient with time or concentration. Alternatively, Bevilacqua and coauthors developed a new analytical formulation for the simulation of the phenomena of diffusion with retention. Its purpose is to reduce all diffusion processes with retention to a unifying phenomenon that can adequately simulate the retention effect. This model may have relevant applications in different areas, such as population spreading with partial hold up of the population to guarantee territorial domain, chemical reactions inducing adsorption processes, and multiphase flow through porous media. In the new formulation, a discrete approach is first formulated with regard to a control parameter that represents the fraction of particles allowed to diffuse, and the governing equation for the modeling of diffusion with retention in a continuum medium requires a fourth order differential term. In order to apply this new formulation to the modeling of practical problems, the newly introduced parameters need to be accurately

D. Campos-Knupp (✉) • L.G. da Silva • A.J. Silva Neto
Mechanical Engineering and Energy Department, Polytechnic Institute, IPRJ-UERJ,
Rua Bonfim, No. 25, Nova Friburgo, RJ, CEP 28625-570, Brazil
e-mail: diegoknupp@iprj.uerj.br; luciano@iprj.uerj.br; ajsneto@iprj.uerj.br

L. Bevilacqua
Universidade Federal do Rio de Janeiro, COPPE-UFRJ, Rio de Janeiro,
RJ, CEP 21941-914, Brazil
e-mail: bevilacqua@coc.ufrj.br

A.C.N.R. Galeão
Laboratório Nacional de Computação Científica, LNCC, Petrópolis,
RJ, CEP 25651-075, Brazil
e-mail: acng@lncc.br

© Springer International Publishing Switzerland 2016
A.J. da Silva Neto et al. (eds.), *Mathematical Modeling and Computational Intelligence in Engineering Applications*, DOI 10.1007/978-3-319-38869-4_7

89

determined through an inverse problem analysis. Hence, this chapter provides an overview of the inverse analysis of anomalous diffusion problems as modeled through this new formulation, and a summary is also presented on the inverse problem formulation and related solution through three different approaches: (1) the maximum likelihood estimation, (2) the Bayesian approach through the *Maximum a Posteriori* objective function, and (3) Markov Chain Monte Carlo methods.

Keywords Anomalous diffusion • Inverse problems • Parameters estimation • Maximum likelihood • Bayesian inference • Maximum a posteriori • Markov chain Monte Carlo

7.1 Introduction

The spread of particles or microorganisms immersed in a given medium or deployed over a given substratum is frequently modeled as a diffusion process, using the well-known diffusion equation derived from the Fick's law. This model represents quite successfully the behavior of several physical phenomena related to dispersion processes; but for some cases, this approach fails to represent the real physical behavior. For instance, the spread of a population of particles may be partially and temporarily blocked when immersed in some particular media; an invading species may hold a fraction of the total population stationed on the conquered territory in order to guarantee territorial domain; and chemical reactions may induce adsorption processes for the diffusion of solutes in liquid solvents in the presence of adsorbent material [4, 5].

Certain physicochemical phenomena need improvement on the analytical formulation due to side effects viewed as unaccountable in the classical diffusion theory. These include the flows though porous media [11]; diffusion processes for some dispersing substances immersed in particular supporting media [1, 2, 6–8]; and the diffusion of hydrogen through metals, a case that is strongly influenced by the presence of hydrogen traps, such as grain boundaries, dislocations, carbides, and non-metallic particles [19]. In most published works on this matter, the well-known second order parabolic equation is assumed as the basic governing equation of the dispersion process, but the anomalous diffusion effect is modeled with the introduction of fractional derivatives [14], the imposition of a convenient variation of the diffusion coefficient with time or concentration [9, 21], or the introduction of complementary terms in the model to account for the retention phenomena [13].

Nevertheless, trying to overcome the anomalous diffusion issue by imposing an artificial dependence by the diffusion coefficient on the particle concentration, or by introducing extra differential terms, while the second order rank of the governing equation is kept unchanged, disguises the real physical phenomenon occurring in the process. In 2011, Bevilacqua and coauthors [4] derived a new analytical formulation to simulate the phenomena of anomalous diffusion. This formulation explicitly takes into account the retention effect in the dispersion process for the purpose of reducing all the diffusion processes with retention to a unifying phenomenon that may adequately simulate the retention effect. In addition to the diffusion coefficient, the newly introduced parameters characterize the blocking process. Specific experimental setups, together with an appropriate inverse analysis, need to be established to determine these complementary parameters.

This chapter investigates an anomalous diffusion inverse problem that does not allow for the simultaneous estimation of all the model parameters. The chapter then presents a characterization procedure in which the true diffusion coefficient is supposedly known; for instance, this information can be supplied by an experimental procedure in which no anomalous diffusion occurs. Hence, the resulting inverse problem consists of a situation in which an anomalous diffusion occurs for the given problem, with the true diffusion coefficient already characterized. The goal is then to estimate the complementary anomalous diffusion parameters. In this situation, it is critically important to consider how the uncertainty present in the supposedly known value of the true diffusion coefficient affects the estimation of the other parameters.

In a general sense, there are two classes of estimation approaches: (1) the classical maximum likelihood procedure, which yields the inverse problem solution as the estimate that maximizes the probability of having a given set of observed data, reducing to the well-known ordinary least squares under certain conditions; and (2) the Bayesian approach, which combines the likelihood function with prior information in order to yield a representation for the parameters posterior distribution. Therefore, in this context, when prior information is available, the Bayesian approach is conceptually the most attractive technique.

In certain conditions, where the experimental error follows a normal distribution and the prior information can be modeled as a normal distribution, it is possible to derive the *Maximum a Posteriori* objective function that may be used to obtain single point estimates for the unknowns, besides the approximation of the confidence intervals. Nevertheless, this represents only part of the information on the unknowns, and the estimated confidence intervals must be used cautiously, since this commonly adopted approach is exact only for linear problems, being just an approximation for nonlinear inverse problems, such as the one discussed herein.

The posterior probability distribution may be further explored using random sampling methods, such as the Markov Chain Monte Carlo (MCMC) techniques that are more computationally intensive; however, these techniques permit to approach the true posterior distribution upon the appropriate modeling of the prior information and the experimental errors, regardless of their statistical distribution form.

This chapter summarizes the inverse problem formulation, as well as the solutions supplied by three different approaches: (1) the classical maximum likelihood estimation, (2) the Bayesian approach through the *Maximum a Posteriori* objective function, and (3) the MCMC methods in order to approximate the posterior probability density distribution. Here, the main goal is to critically address the different aspects of such methodologies for the anomalous diffusion inverse problem under study as an important step toward the analysis of real problems using the new model.

7.2 Direct Problem Formulation and Solution

Consider the process that is schematically represented in Fig. 7.1. The redistribution of the contents of each cell indicates that a fraction of the contents αp_n is retained in the nth cell and the exceeding volume is evenly transferred to the neighboring cells, that is, $0.5\beta p_n$ to the left, to the $(n-1)$th cell and $0.5\beta p_n$ to the right, to the $(n+1)$th cell, at each time step, where $\beta = 1 - \alpha$. This means that the dispersion runs more slowly than for the classical diffusion problem. Note that if $\beta = 1$, the problem is reduced to the classical distribution.

This process can be written as the following algebraic expressions:

$$p_n^t = (1 - \beta)p_n^{t-1} + \frac{1}{2}\beta p_{n-1}^{t-1} + \frac{1}{2}\beta p_{n+1}^{t-1} \tag{7.1a}$$

$$p_n^{t+1} = (1 - \beta)p_n^t + \frac{1}{2}\beta p_{n-1}^t + \frac{1}{2}\beta p_{n+1}^t \tag{7.1b}$$

Manipulating (7.1) in order to obtain finite difference terms yields

$$\frac{\Delta p_n^{t+\Delta t}}{\Delta t} = \beta \left\{ \frac{1}{2}\frac{L_0^2}{T_0}\frac{\Delta^2 p_n}{\Delta x^2} + \frac{O\left(\Delta x^2\right)}{\Delta x^2} - (1-\beta)\frac{1}{4}\frac{L_1^4}{T_0}\frac{\Delta^4 p_n}{\Delta x^4} \right\}^{t-\Delta t} \tag{7.2}$$

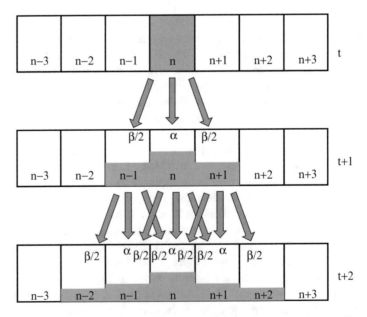

Fig. 7.1 Schematic representation of the symmetric distribution with retention $\alpha = (\beta - 1)$

where T_0, L_0, and L_1 are integration parameters, described with more details in [4]. Calling $K_2 = L_0^2/2T_0$ and $K_4 = L_1^4/4T_0$, both considered constant in this chapter, and taking the limit as $\Delta x \to 0$ and $\Delta t \to 0$, the following applies:

$$\frac{\partial p(x,t)}{\partial t} = \beta K_2 \frac{\partial^2 p(x,t)}{\partial x^2} - \beta(1-\beta) K_4 \frac{\partial^4 p(x,t)}{\partial x^4} \tag{7.3a}$$

The fourth order term with negative sign introduces the anomalous diffusion effect that shows up naturally, without any artificial assumption, as an immediate consequence of the temporary retention imposed by the redistribution law. Further discussion on the model derivation can be found in reference [4].

As the test case for the present chapter, consider the governing equation given by (7.3a), valid for $0 < x < 1$ and $t > 0$, with the following boundary and initial conditions:

$$p(0,t) = 1, \quad p(1,t) = 1, \quad \left.\frac{\partial p(x,t)}{\partial x}\right|_{x=0} = 0, \quad \left.\frac{\partial p(x,t)}{\partial x}\right|_{x=1} = 0, \quad t > 0 \tag{7.3b}$$

$$p(x,0) = f(x) = 2\sin^{100}(\pi x) + 1, \quad 0 \le x \le 1 \tag{7.3c}$$

The problem given by (7.3) is solved here using the *NDSolve* routine of the *Mathematica*® system [20], under automatic absolute and relative error control. In regard to the inverse problem solution, from an observation of the problem defined in (7.3), it is evident that the three parameters appearing in the model cannot be simultaneously estimated in the absence of prior information, since three parameters are defining two coefficients in (7.3a); i.e., there are infinite sets of values for the parameters $\mathbf{Z} = \{\beta, K_2, K_4\}$ that lead to the exactly same mathematical formulation, yielding non-uniqueness of the inverse problem solution, as already illustrated by the sensitivity analysis in [17]. Since the most interesting aspect of the problem investigated is the identification of the three parameters appearing in the model, due to their direct physical interpretation [4], a decision was made not to rewrite the problem in terms of two coefficients (which would be multiplying the second and fourth order differential terms). Instead, it is hereby considered that prior information can be obtained for the true diffusion coefficient, K_2. This prior information could be obtained through an independent experiment; for example, an inverse problem in a physical situation where the blocking process that characterizes the anomalous diffusion phenomenon does not occur. The inverse problem formulation and solution approaches employed herein to tackle this problem are addressed in the following section.

7.3 Inverse Problem

In order to investigate the inverse problem solution concerning the estimation of the model parameters $\mathbf{Z}$, a set of experimental data is considered available, $\mathbf{p}^{exp}$, which are simulated in this chapter with the solution of (7.3) in which noise is added from a normal distribution with known variance:

$$p_i^{exp} = p_i \left(\mathbf{Z}_{exact} \right) + \epsilon_i, \quad \epsilon \sim N \left(0, \sigma_e^2 \right) \tag{7.4}$$

The next sections present the inverse problem formulation and solution by means of the maximum likelihood and Bayesian approaches.

7.3.1 Maximum Likelihood

Assuming that the measurement errors related to the data $\mathbf{p}^{exp}$ are additive, uncorrelated, and have normal distribution, then the probability density for the occurrence of the measurements $\mathbf{p}^{exp}$ with the given parameters values $\mathbf{Z}$ may be expressed as [10]

$$\pi \left(\mathbf{p}^{exp} | \mathbf{Z} \right) = (2\pi)^{-N_d/2} \, |\mathbf{W}|^{-1/2}$$
$$exp \left\{ -\frac{1}{2} \left[\mathbf{p}^{exp} - \mathbf{p}^{calc} \left(\mathbf{Z} \right) \right]^T \mathbf{W}^{-1} \left[\mathbf{p}^{exp} - \mathbf{p}^{calc} \left(\mathbf{Z} \right) \right] \right\} \tag{7.5}$$

where N_d is the number of experimental data employed and $\mathbf{p}^{calc}$ is the vector containing the quantities calculated through the direct problem solution employing parameters values $\mathbf{Z}$. Hence, the likelihood estimates can be seen as the values of $\mathbf{Z}$ that maximize the likelihood function given by (7.5), which may be achieved with the minimization of the argument of the exponential function in (7.5). Assuming that the measurements variance is constant, then it is equivalent to minimize

$$Q_{ML} \left(\mathbf{Z} \right) = \left[\mathbf{p}^{exp} - \mathbf{p}^{calc} \left(\mathbf{Z} \right) \right]^T \left[\mathbf{p}^{exp} - \mathbf{p}^{calc} \left(\mathbf{Z} \right) \right] = \sum_{i=1}^{N_d} \left[p_i^{exp} - p_i^{calc} \left(\mathbf{Z} \right) \right]^2 \tag{7.6}$$

In the hypotheses assumed herein, the minimization of the least-squares norm given by (7.6) yields maximum likelihood estimates. In this chapter, for this purpose, the Levenberg–Marquardt iterative procedure for nonlinear problems [12] is employed:

$$\mathbf{Z}^{n+1} = \mathbf{Z}^n + \left[\left(\mathbf{J}^T \right)^n \mathbf{J}^n + \lambda^n \mathbf{I} \right]^{-1} \left(\mathbf{J}^T \right)^n \left[\mathbf{p}^{exp} - \mathbf{p}^{calc} \left(\mathbf{Z}^n \right) \right] \tag{7.7}$$

where $\mathbf{I}$ is the identity matrix, λ is the damping factor, which is adjusted at each iteration and becomes close to zero when the convergence has been achieved. The elements of the Jacobian matrix $\mathbf{J}$ are given by

$$J_{ij} = \frac{\partial p_i^{calc}}{\partial Z_j}, \qquad i = 1, 2, \ldots, N_d, \ \ j = 1, 2, \ldots, N_{un} \tag{7.8}$$

where N_{un} is the number of unknowns being estimated in the inverse problem solution.

The standard deviation of the estimated parameters as result of the uncertainty in the experimental data may be obtained through the following expression [3]:

$$\sigma_{Z_i} = \sqrt{\left[\left(\mathbf{J}^T \mathbf{W}^{-1} \mathbf{J} \right)^{-1} \right]_{i,i}} \qquad i = 1, 2, \ldots, N_{un} \tag{7.9}$$

which is only exact for linear inverse problems, but it is commonly adopted as an approximation for nonlinear inverse problems, such as the problem discussed herein. For instance, the confidence intervals for a 95 % confidence degree are approximately calculated as:

$$Z_i - 2\sigma_{Z_i} < Z_i < Z_i + 2\sigma_{Z_i}, \qquad i = 1, 2, \ldots N_{un} \tag{7.10}$$

7.3.2 Bayesian Inference

In the statistical inversion theory; namely, the Bayesian approach, the inverse problem is formulated as a problem of statistical inference and it is based on the following principles [10]: (1) All the variables in the model are modeled as random variables; (2) The randomness describes the degree of information available; (3) The degree of information is coded in probability distributions; and (4) The solution of the inverse problem is the posterior probability distribution. Thus, in the Bayesian approach, all the possible information is incorporated in the model, thus yielding a better uncertainty assessment of the estimated parameters.

Consider that prior information about the parameters $\mathbf{Z}$ is available. Assuming that this information can be modeled as a probability density $\pi_{pr} (\mathbf{Z})$, then the Bayes' theorem of inverse problems can be expressed as [10]

$$\pi_{post} (\mathbf{Z}) = \pi (\mathbf{Z}| \mathbf{p}^{exp}) = \frac{\pi_{pr} (\mathbf{Z}) \, \pi (\mathbf{p}^{exp}| \mathbf{Z})}{\pi (\mathbf{p}^{exp})} \tag{7.11}$$

where $\pi_{post} (\mathbf{Z})$ is the posterior probability density, $\pi_{pr} (\mathbf{Z})$ is the prior information on the unknowns, modeled as a probability distribution, $\pi (\mathbf{p}^{exp}| \mathbf{Z})$ is the likelihood function, already defined in (7.5) for the case of normally distributed errors, and $\pi (\mathbf{p}^{exp})$ is the marginal density and acts as a normalizing constant.

This methodology produces a distribution that may be explored in various ways and using different methods. One of the most common approaches to obtain estimates is through the *maximum a posteriori* (MAP) estimator that may be used to produce single point estimates for the unknowns, provided that the prior information can be modeled as a normal distribution, as well as to obtain approximated confidence intervals for the parameters posterior distribution. Nonetheless, within the statistical framework, the point estimates represent only part of the information on the unknowns, and the approximated confidence intervals may not be very accurate for highly nonlinear problems. Alternatively, the posterior probability distribution may be explored through random sampling methods, such as the MCMC techniques [10, 15]. Both the MAP estimator and the MCMC are briefly described next.

7.3.2.1 Maximum a Posteriori

Consider that the prior information on parameters $\mathbf{Z}$ can be modeled as a normal distribution. Thus, $\pi_{\text{pr}}(\mathbf{Z})$ can be expressed by

$$\pi_{\text{pr}}(\mathbf{Z}) = (2\pi)^{-N_{\text{un}}/2} |\mathbf{V}|^{-1/2} \exp\left[-\frac{1}{2}(\mathbf{Z} - \boldsymbol{\mu}_{\text{pr}})^T \mathbf{V}^{-1}(\mathbf{Z} - \boldsymbol{\mu}_{\text{pr}})\right] \qquad (7.12)$$

where $\mathbf{V}$ and $\boldsymbol{\mu}_{\text{pr}}$ are, respectively, the covariance matrix and the mean of the assumed prior information for $\mathbf{Z}$. Substituting (7.5) and (7.12) into (7.11), the following is obtained:

$$\ln\left[\pi_{\text{post}}(\mathbf{Z})\right] \propto$$
$$-\frac{1}{2}\left[(N_{\text{un}} + N_d)\ln 2\pi + \ln\left|\mathbf{W}^{-1}\right| + \ln\left|\mathbf{V}^{-1}\right| + Q_{\text{MAP}}(\mathbf{Z})\right] \qquad (7.13)$$

where

$$Q_{\text{MAP}}(\mathbf{Z}) = \left[\mathbf{p}^{\text{exp}} - \mathbf{p}^{\text{calc}}(\mathbf{Z})\right]^T \mathbf{W}^{-1}\left[\mathbf{p}^{\text{exp}} - \mathbf{p}^{\text{calc}}(\mathbf{Z})\right]$$
$$+ (\mathbf{Z} - \boldsymbol{\mu}_{\text{pr}})^T \mathbf{V}^{-1}(\mathbf{Z} - \boldsymbol{\mu}_{\text{pr}}) \qquad (7.14)$$

is known as the maximum a posteriori (MAP) objective function. The minimization of $Q_{\text{MAP}}(\mathbf{Z})$ yields estimates $\mathbf{Z}$ which maximize the posterior distribution $\pi_{\text{post}}(\mathbf{Z})$. In this chapter, the MAP objective function is minimized with the iterative procedure of the Gauss–Newton method [3]

$$\mathbf{Z}^{n+1} = \mathbf{Z}^n + \left[(\mathbf{J}^T)^n \mathbf{W}^{-1}\mathbf{J}^n + \mathbf{V}^{-1}\right]^{-1}$$
$$\times \left[(\mathbf{J}^T)^n \mathbf{W}^{-1}\left[\mathbf{p}^{\text{exp}} - \mathbf{p}^{\text{calc}}(\mathbf{Z})\right] + \mathbf{V}^{-1}(\mathbf{Z}^n - \boldsymbol{\mu}_{\text{pr}})\right] \qquad (7.15)$$

where the Jacobian matrix elements have already been defined in (7.8). Similarly to the case of the maximum likelihood estimates, the standard deviation for the estimated parameters, which is an approximation of the standard deviation of the posterior distribution and is supposed to follow a normal distribution in this approximation, can be obtained through the following expression:

$$
\sigma_{Z_i} = \sqrt{\left[\left(\mathbf{J}^T \mathbf{W}^{-1} \mathbf{J} + \mathbf{V}^{-1}\right)^{-1}\right]_{i,i}}, \qquad i = 1, 2, \ldots, N_{\text{un}} \tag{7.16}
$$

As mentioned above, although frequently used as an approximation for nonlinear problems, the expression given in (7.16) only provides an exact representation for linear problems. Even in the case where the experimental error and the prior distributions follow the normal distribution, the parameters distributions do not necessarily follow the normal distribution in the case of nonlinear models, as discussed in [16].

7.3.2.2 Markov Chain Monte Carlo

When the corresponding posterior distributions may not be obtained, a simulation-based method is needed. The inference based on simulation techniques uses samples from the posterior distribution in order to approximately construct this density and extract information from it. Several sampling strategies are proposed in the literature, including the Monte Carlo method via Markov Chain (MCMC) [10, 15], which is widely employed and is adopted in this chapter.

In order to implement the Markov Chain a candidate-generating density is needed, $q\left(\mathbf{Z}^t, \mathbf{Z}^*\right)$, which denotes a source density for a candidate draw $\mathbf{Z}^*$, given the current state $\mathbf{Z}^t$. Then, the Metropolis–Hastings algorithm [10, 15] is used herein to implement the MCMC method. The procedure can be briefly defined by the following steps:

Step 1: Sample a candidate $\mathbf{Z}^*$ from the candidate-generating density $q\left(\mathbf{Z}^t, \mathbf{Z}^*\right)$

Step 2: Calculate

$$
\alpha = \min\left[1, \frac{\pi\left(\mathbf{Z}^* \mid \mathbf{P}^{\text{exp}}\right) q\left(\mathbf{Z}^*, \mathbf{Z}^t\right)}{\pi\left(\mathbf{Z}^t \mid \mathbf{P}^{\text{exp}}\right) q\left(\mathbf{Z}^t, \mathbf{Z}^*\right)}\right] \tag{7.17a}
$$

Step 3: If $U\left(0, 1\right) < \alpha$, then

$$
\mathbf{Z}^{t+1} = \mathbf{Z}^* \tag{7.17b}
$$

else

$$
\mathbf{Z}^{t+1} = \mathbf{Z}^t \tag{7.17c}
$$

where $U\left(0, 1\right)$ is a random number from a uniform distribution between 0 and 1.

Step 4: Return to **Step 1** in order to generate the chain $\{\mathbf{Z}^1, \mathbf{Z}^2, \ldots, \mathbf{Z}^{N_{MCMC}}\}$. The first states of this chain must be discarded until the convergence of the chain is reached. These ignored samples are called the burn-in period, whose length will be denoted by $N_{\text{burn−in}}$.

In the present chapter, a random walk process was used in order to generate the candidates, so that $\mathbf{Z}^* = \mathbf{Z}^t + \boldsymbol{\eta}$, where $\boldsymbol{\eta}$ follows the distribution q, which was defined as a normal density. In this case, the distribution is symmetric, therefore $q(\mathbf{Z}^*, \mathbf{Z}^t) = q(\mathbf{Z}^t, \mathbf{Z}^*)$, consequently Step 2 is simplified, and Eq. (7.17a) may be rewritten as

$$\alpha = \min\left[1, \frac{\pi\left(\mathbf{Z}^* \mid \mathbf{P}^{\text{exp}}\right)}{\pi\left(\mathbf{Z}^t \mid \mathbf{P}^{\text{exp}}\right)}\right] \tag{7.17d}$$

7.4   Results and Discussion

From observing Eq. (7.3a) it is clear that the three parameters appearing in the model, i.e., β, K_2, and K_4, cannot be simultaneously estimated in the absence of prior information, since these parameters define only two coefficients. Therefore, the results presented below consider that prior information is available for the true diffusion coefficient, K_2, which may, in a physical real problem, be obtained through an independent experimental procedure; for instance, a situation for which the given problem is known to lack any anomalous diffusion effect (in this case K_2 would be the only parameter to be estimated). It is also assumed that the prior information on K_2 may be modeled as a normal distribution with 1.0×10^{-3} mean and 0.1×10^{-3} standard deviation. The experimental data have been simulated for transient measurements of p at $x = 0.5$, from $t = 5$ up to $t = 95$, employing the values $\beta = 0.2$, $K_2 = 1.0 \times 10^{-3}$, and $K_4 = 1.0 \times 10^{-5}$ and $\sigma_e = 0.02$ in Eq. (7.4), yielding up to 4 % error in the experimental data employed. Figure 7.2 illustrates the set of experimental data employed, together with the curve obtained from the solution of problem (7.3) for the test case under study. This case has been investigated before in [17, 18], and its sensitivity analysis indicates that $x = 0.5$ is the most appropriate choice for the sensor location.

For the maximum likelihood approach the prior information is not incorporated into the model, and commonly the *a-priori* characterized parameter is assumed to be exactly known. Alternatively, if an accurate assessment of the uncertainty in the estimated parameters is needed, error propagation techniques can be employed [18]. Here, the inverse problem solution with the maximum likelihood approach has been carried out regarding three different fixed values for K_2: 1.0×10^{-3}, 0.8×10^{-3}, and 1.2×10^{-3}, which correspond to the mean value, the inferior, and the superior 95 % confidence bounds of the prior distribution, respectively. Those results are presented in Table 7.1. If the lowest and highest bounds for the estimates of β and K_4 are established, then it may be inferred from those results that β should be in the interval $[0.134, 0.257]$ and K_4 in the interval $\left[0.719 \times 10^{-5}, 1.601 \times 10^{-5}\right]$.

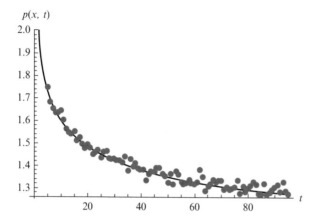

Fig. 7.2 Simulated experimental data (*red dots*) for transient measurements of a sensor located at $x = 0.5$. The *black curve* shows the numerical solution employed to simulate the experimental data

Table 7.1 Estimates obtained with the maximum likelihood approach employing three different fixed values for K_2

Estimated mean and 95 % confidence intervals			
Parameter	$K_2 = 1.0 \times 10^{-3}$	$K_2 = 0.8 \times 10^{-3}$	$K_2 = 1.2 \times 10^{-3}$
β	0.138[0.161, 0.205]	0.229[0.202, 0.257]	0.153[0.134, 0.171]
$K_4 \left[\times 10^{-5}\right]$	1.111[0.838, 1.384]	0.942[0.719, 1.166]	1.286[0.964, 1.601]

For the purpose of the inverse problem solution through the Bayesian framework, some aspects of the Markov chains simulated with the MCMC approach are first illustrated. Figure 7.3a–c presents the evolution of the chains related to β, K_2, and K_4, respectively, which shows that after 5,000 states all chains are already converged to the equilibrium distribution. This fact is also depicted in Fig. 7.3d, where the autocorrelation functions of those three chains are plotted. To be clear, the solution using the Bayesian framework does not require that a fixed value be set for K_2; instead, the prior distribution itself is incorporated into the model. The MCMC states presented in Fig. 7.3b allude to samples of the posterior distribution of K_2, which is not necessarily the same as the prior distribution.

Figure 7.4a–c describes the parameters distributions provided by the three approaches discussed herein, for β, K_2, and K_4, respectively. The distribution resulting from the MCMC method is an approximation of the posterior distribution samples, as generated from the Markov chains after the burn-in states (the first 5,000 states, as mentioned above) are discarded. The distribution provided by the maximum a posteriori estimator is a normal distribution, with the mean given by the estimates and the standard deviation calculated using the expression given by (7.16). It must be reminded that this is an approximation for the problem discussed herein, since it involves a nonlinear model and the parameters distributions do

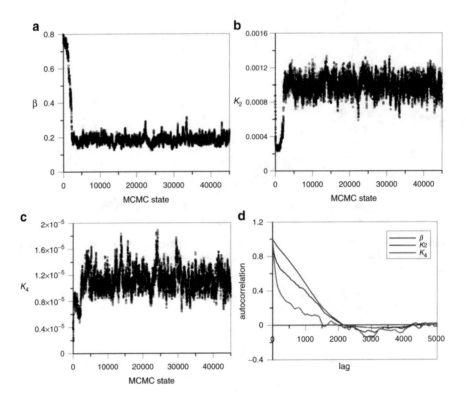

Fig. 7.3 Evolution of the Markov chains related to (**a**) β, (**b**) K_2, (**c**) K_4. (**d**) Autocorrelation function with respect to the three chains

not necessarily follow a normal distribution. The distributions presented for the maximum likelihood approach are normal distributions with the mean given by the estimates and the standard deviation calculated using Eq. (7.9), obtained with an imposed fixed value of $K_2 = 1.0 \times 10^{-3}$. These distributions are presented only as a reference, since they do not supply any error information due to uncertainty over the employed value of K_2.

Figure 7.4a shows that the posterior distributions for K_2 obtained with the MCMC and the MAP estimator are very similar, and the deviations seem to be only due to the finite number of samples in the MCMC, which clearly tends to a normal distribution. Those distributions are evidently wider than the reference maximum likelihood distribution, thus providing deep insight into the uncertainty of the estimated value of β as a result of the prior information associated with K_2. Hence, it is very clear that choosing a fixed value for K_2 and implementing the classical maximum likelihood approach, without any error propagation approach, leads to quite unrealistic confidence intervals in this case.

Figure 7.4b illustrates good consistency between the posterior distributions obtained through the MAP and MCMC approaches. This figure also plots the prior

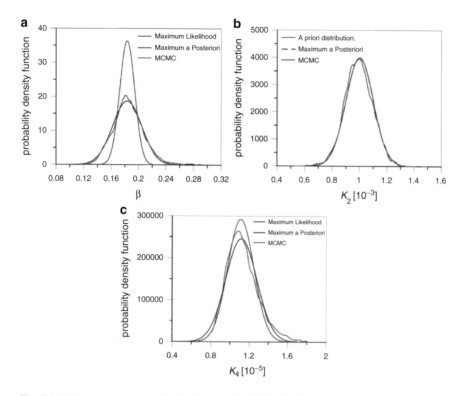

Fig. 7.4 Estimated parameters distributions: (**a**) β; (**b**) K_2; (**c**) K_4

distribution, that turns out to essentially match the posterior distribution. In fact, this outcome was expected, since this inverse problem involves the estimation of three parameters that define only two coefficients in the direct problem formulation, (7.3a); in other words, the experimental measurements do not supply any additional information concerning this parameter.

Figure 7.4c shows that the posterior distribution for K_4 obtained with the MCMC is a little skewed, and does not seem to approach the normal distribution. In this case, even though the MAP distribution still offers a fairly good approximation, the effects of the nonlinearity of the model are quite clear, indicating that for more accurate assessments of the posterior distribution the MCMC methodology should be preferred. It is also interesting to notice that the maximum likelihood distribution is much closer to the posterior distribution in Fig. 7.4c than in Fig. 7.4a, suggesting that the uncertainty over the prior information with respect to K_2 is much more sensitive for β than for K_4. This is directly related to the fact that K_2 appears multiplied by β in the second order differential term in (7.3a), whereas K_2 does not appear in the fourth order differential term in (7.3a), where K_4 is present.

The posterior distribution behavior may be better observed in Fig. 7.5, which depicts the true joint posterior distribution samples for β and K_4 from the MCMC,

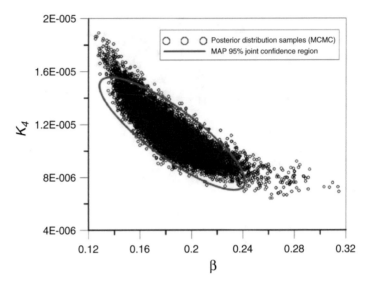

Fig. 7.5 Joint posterior distribution of β and K_4

Table 7.2 Summary of the estimates obtained

Parameter	Maximum a posteriori	MCMC	Maximum likelihood[a]
β	0.183[0.140, 0.226]	0.186[0.145, 0.234]	[0.134, 0.257]
$K_2\left[\times 10^{-3}\right]$	1.0[0.8, 1.2]	0.99[0.79, 1.19]	[0.8, 1.2][b]
$K_4\left[\times 10^{-5}\right]$	1.117[0.793, 1.442]	1.127[0.85, 1.52]	[0.719, 1.601]

[a]Refers to the lowest and highest bounds presented in Table 7.1
[b]Refers to the 95 % confidence interval of the prior distribution

together with the 95 % joint elliptic region obtained from the approximated normal distribution estimated with the MAP estimator. The true posterior distribution region is non-convex, and clearly deviates from the elliptic region. Nevertheless, the relative merits of the approximated region should be highlighted as a fairly good approximation, with much lower computational effort.

Finally, Table 7.2 summarizes the posterior distribution estimates for β, K_2, and K_4, together with the 95 % degree confidence intervals, as obtained with the MAP estimator and the MCMC method. The lowest and highest bounds supplied by the maximum likelihood approach (initially presented in Table 7.1) are also shown as reference. The results indicate that both the MAP and MCMC approaches yield reliable confidence intervals for the parameters. The main advantage of the MCMC is its ability to recover the true posterior distribution. The MAP approach also produced fairly good approximations with a much less intensive computational effort.

7.5 Conclusions

This chapter summarizes three different approaches for the formulation and solution of inverse anomalous diffusion problems: the classical maximum likelihood approach and two options within the Bayesian framework: the *maximum a posteriori* estimator and the MCMC method. The problem under analysis requires the availability of prior information, and under these circumstances, the Bayesian approach is more attractive, since it allows for the incorporation of the prior information into the model, as demonstrated herein. The results described in this chapter show that the parameters posterior distribution does not seem to follow a normal distribution, as illustrated by the MCMC results. Nevertheless, the MAP approach has relative merits: it yields reliable approximated confidence intervals for the parameters posterior distributions with much less computational effort, thanks to the fact that the number of runs of the direct problem solution needed is much lower in comparison with the MCMC approach.

Acknowledgements The authors acknowledge the financial support provided by the Brazilian Agencies FAPERJ, Fundação Carlos Chagas Filho de Amparo à Pesquisa do Estado do Rio de Janeiro; CNPq, Conselho Nacional de Desenvolvimento Científico e Tecnológico; and CAPES, Coordenação de Aperfeiçoamento de Pessoal de Nível Superior.

References

1. Ashmawy, A., Muhammad, N., Elhajji, D.: Advection, diffusion, and sorption characteristics of inorganic chemicals in GCL bentonite. In: Waste Containment and Remediation, American Society of Civil Engineers Library, ASCE. pp. 1–10. American Society of Civil Engineers Library, ASCE (2005). doi: 10.1061/40789(168)4
2. Atsumi, H.: Hydrogen bulk retention in graphite and kinetics of diffusion. J. Nucl. Mater. **307/311**, 1466–1470 (2002)
3. Beck, J., Arnold, K.: Parameter Estimation in Engineering and Science. Wiley Interscience, New York (1977)
4. Bevilacqua, L., Galeão, A.C.N.R., Costa, F.P.: A new analytical formulation of retention effects on particle diffusion process. Ann. Braz. Acad. Sci. **83**, 1443–1464 (2011)
5. Brandani, S., Jama, M., Ruthven, D.: Diffusion, self-diffusion and counter-diffusion of benzene and p-xylene in silicalite. Microporous Mesoporous Mater. **35/36**, 283–300 (2000)
6. D'Angelo, M., Fontana, E., Chertcoff, R., Rose, M.: Retention phenomena in non-Newtonian fluid flows. Phys. A **327**, 44–48 (2003)
7. Deleersnijder, E., Beckers, J., Delhez, E.: The residence time of setting in the surface mixed layer. Environ. Fluid Mech. **6**, 25–42 (2006)
8. Green, P.: Translational dynamics of macromolecules in metals. In: Neogi, P. (ed.) Diffusion in Polymers. CRC, Boca Raton (1996)
9. Joannès, S., Mazé, L., Bunsell, A.: A concentration-dependent diffusion coefficient model for water sorption in composite. Compos. Struct. **108**, 111–118 (2014)
10. Kaipio, J., Somersalo, E.: Statistical and Computational Inverse Problems, vol. 160. Springer, New York (2005)
11. Liu, H., Thompson, K.: Numerical modeling of reactive polymer flow in porous media. Comput. Chem. Eng. **26**, 1595–1610 (2002)

12. Marquardt, D.W.: An algorithm for least-squares estimation of non-linear parameters. J. Soc. Ind. Appl. Math. **11**, 431–441 (1963)
13. McNabb, A., Foster, P.: A new analysis of the diffusion of hydrogen in iron and ferritic steels. Trans. Metall. Soc. AIME **227**, 618–1963 (1963)
14. Metzler, R., Klafter, J.: The random walk's guide to anomalous diffusion: a fractional dynamics approach. Phys. Rep. **339**, 1–77 (2000)
15. Paez, M.: Bayesian approaches for the solution of inverse problems. Thermal Measurements and Inverse Techniques, pp. 437–455. CRC, Boca Raton (2011)
16. Schwaab, M., Biscaia, Jr. E.C., Monteiro, J.L., Pinto, J.: Nonlinear parameter estimation through particle swarm optimization. Chem. Eng. Sci. **63**, 1542–1552 (2008)
17. Silva, L.G., Knupp, D.C., Bevilacqua, L., Galeão, A.C.N.R., Simas, J.G., Vasconcellos, J.F., Silva Neto, A.J.: Investigation of a new model for anomalous diffusion phenomena by means of an inverse analysis. In: Proc 4th Inverse Problems, Design and Optimization Symposium. Albi, France (2013)
18. Silva, L.G., Knupp, D.C., Bevilacqua, L., Galeão, A.C.N.R., Silva Neto, A.J.: Inverse problem in anomalous diffusion with uncertainty propagation. Comput. Assisted Methods Eng. Sci. **21**, 245–255 (2014)
19. Turnbull, A., Carroll, M., Ferriss, D.: Analysis of hydrogen diffusion and trapping in a 13% chromium martensitic stainless steel. Acta Metall. **37**(7), 2039–2046 (1989)
20. Wolfram, S.: The Mathematica Book. Wolfram Media, Cambridge (2005)
21. Wu, J., Berland, K.: Propagators and time-dependent diffusion coefficients for anomalous diffusion. Biophys. J. **95**, 2049–2052 (2008)

Chapter 8
Accelerated Direct-Problem Solution: A Complementary Method for Computational Time Reduction

Alberto Prieto-Moreno, Leôncio D. Tavares Câmara, and Orestes Llanes-Santiago

Abstract This paper presents a proposal designed to reduce the time required by the process to estimate the parameters of a system by accelerating the direct-problem solution as the slow phase in any estimation method. This proposal is considered a complement to existing procedures, such as the combination of different optimization methods for the purpose of reducing the number of calls to the objective function. The proposal consists of a procedure that helps study the relation between the direct-problem solution step and the time required for this solution, as well as the influence of the direct solution's built-in error on the accuracy of the estimated parameters. Consequently, the extent in which the estimation process can be accelerated without impairing estimation accuracy can be determined. For the purpose of testing its viability, this proposal was applied to the estimation problem of the kinetic parameters of a chromatography column process, as modeled using the front-velocity method. The results from this test show that, by accelerating the direct-problem solution, the estimation time can be reduced significantly, without affecting the accuracy of the estimation.

Keywords Parameter estimation • Accuracy • Computational cost • Direct solution • Chromatographic column • Front velocity • Kinetic parameters

A. Prieto-Moreno (✉) • O. Llanes-Santiago
Automatic and Computing Department, Instituto Superior Politécnico José Antonio Echeverría (CUJAE), Marianao, La Habana, CP 19390, Cuba
e-mail: albprieto@electrica.cujae.edu.cu; orestes@electrica.cujae.edu.cu

L.D.T. Câmara
Mechanical Engineering and Energy Department, Polytechnic Institute, IPRJ-UERJ, Rua Bonfim, No. 25, Nova Friburgo, RJ, CEP 28625-570, Brazil
e-mail: dcamara@iprj.uerj.br

© Springer International Publishing Switzerland 2016
A.J. da Silva Neto et al. (eds.), *Mathematical Modeling and Computational Intelligence in Engineering Applications*, DOI 10.1007/978-3-319-38869-4_8

8.1   Introduction

For the purpose of identifying the optimal parameters in an optimization problem, numerical techniques are used. These techniques can be classified into two large groups: deterministic and stochastic methods.

In their search for parameters, the deterministic methods generally rely on information on the gradient of the function to be minimized, hence determining the direction where the problem solution can be found. These methods include, *inter alia*, the Newton's and Levengberg–Marquardt methods [15]. The primary advantage of these methods is their fast convergence to a stationary point; however, these methods do not guarantee that a global minimum (or maximum) point will be achieved. Furthermore, they need to work with derivatives that can be quite expensive.

For their part, given their random characteristics, the stochastic methods can offer probabilistic assurances that the global minimum will be reached. These methods have proven to be more efficient than their traditional deterministic counterparts [20], but their high computational cost poses an inconvenience.

Where computational techniques are applied, it is always an objective to cut down their cost by either reducing the number of operations required or cutting down their operating time [11]. In their application to practical problems with time restriction, the global stochastic-based optimization techniques have the disadvantage of delayed generation of a solution. As a result, its application to problems that require on-line decision-making is restricted.

The stochastic optimization methods are based on a large variety of functioning principles. Consequently, their computational cost is often measured on basis of the number of calls to the objective function required by the method [9]. In order to mitigate the computational cost problem, many authors have proposed hybrid methods that combine stochastic and deterministic techniques [3, 13, 19]. These methods primarily seek a smaller number of calls to the objective function by relying on techniques that rapidly reduce the search space, combined with techniques that have demonstrated fast convergence.

These methods need the solution of models expressed in differential equations. With very few exceptions, many of the differential equations of interest stem from describing or modeling natural phenomena or other real-life problems. These equation systems not always have an analytical solution, and therefore, they cannot be solved accurately. Instead, these solutions can be approximated using numerical methods, including but not limited to Euler, Runge Kutta, Runge Kutta Fehlberg, and Adams–Bashforth methods [4]. Such methods apply different principles in order to make sure that the solution of the model is more accurate.

This chapter presents a different approach that seeks to cut down the estimation time by reducing the direct-problem solution time in each call to the objective function. For this purpose, it is hereby proposed that the accuracy of the direct-problem solution be reduced in order to accelerate the solution procedure. This proposal is based on the hypothesis that the optimization method will be able to produce good estimates, even if the error in the direct-problem solution increases with respect to experimental data.

The purpose of this chapter is three-fold; namely, (1) reduce the direct-problem solution time; (2) assess the influence exercised by the increased error that is built in the direct-problem solution on the estimation accuracy; and (3) propose a procedure that helps determine the maximum error that may be introduced into the direct-problem solution, in a manner that maintains the accuracy of the solution supplied by the optimization method.

The paper has the following structure: Sect. 8.2 describes the chromatography column modeling technique, known as front velocity, being the real process that will be used to illustrate the proposal contained herein. Section 8.3 describes the procedure applied to determine the extent in which the direct-problem solution may be accelerated without affecting the solution of the inverse problem. The results and their discussion are presented in Sect. 8.4. Finally, conclusions are presented.

8.2 Materials and Methods

The chromatography column model determines the performance of a simulated moving bed (SMB) in its final separation, depending on the number of interconnected columns involved. In general, different research teams have used dispersion models [10] to represent the chromatography column. These models are robust and efficient, but require a deep numerical treatment of partial differential equations that entails high computational costs.

This paper applies a new modeling approach, known as convection front velocity [2, 5]. In the front-velocity modeling technique, the convection of the liquid phase is considered the main phenomenon of the transportation of molecules through the chromatographic column, followed by the mass transfer between the solid adsorbent and the liquid phase.

This modeling technique can be used as a powerful tool to determine the chromatographic behavior of a sample, since it can be easily implemented for routine analysis and requires a smaller number of parameters.

8.2.1 Front Velocity

Since the flow rate within the column is determined by an external pumping system, the time that the liquid phase needs to travel the full chromatographic column can be established, provided that the volumetric flow rate, the porosity, and the column volume are known experimentally. In the chromatographic column shown in Fig. 8.1, the control volume of size J^* moves along the column at the same speed as the eluent flow. For this case, the column length is discretized with control volumes of size J^*.

The time interval (Δt) in which the liquid phase moves during each control volume is obtained using the expression

$$\Delta t = \frac{\varepsilon V}{nF} \tag{8.1}$$

Fig. 8.1 Chromatographic
column of length L with a
volume discretization volume
of size J^* [5]

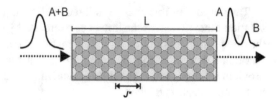

where ε, V, n, and F represent the bed porosity of the column, the total volume of the column, the total number of control volumes, and the liquid flow rate, respectively.

For the purpose of modeling the mass transfer, two concentrated mass-transfer models are assumed, as described by Eqs. (8.2) and (8.3), where C, q, k_1 and k_2 are, respectively, the concentration in the liquid and solid phases, and the global absorption and desorption constants of mass-transfer kinetics.

$$\frac{dC}{dt} = -k_1 C + k_2 q \tag{8.2}$$

$$\frac{dq}{dt} = -\frac{dC}{dt} \tag{8.3}$$

8.2.2 Direct Solution

A large number of engineering problems are formulated as a differential equation or a system of differential equations. A typical problem consists in finding the solution for a set of differential equations that are subject to a set of initial values. For the differential equations in general, and especially for nonlinear differential equations, analytical solutions cannot often be found. For this reason, numerical solutions are required. As a result, a set of values (or variables) of the dependent variable as a function of the independent variable (or variables) is obtained in a finite number of points.

The equations that describe the mass-transfer dynamics are first-order differential equations. The numerical procedure for the solution of these equations is shown graphically in Fig. 8.2.

The solid curve is the solution of some differential equation. The purpose here is to obtain a numerical approximation for the solid curve in a set of finite points identified in the figure as $t_{n-1}, t_n, t_{n+1}, t_{n+2}$. The solutions in the relevant points are identified as $y_{n-1}, y_n, y_{n+1}, y_{n+2}$, where n represents a certain time instant. If the solution at a certain point y_n and t_n is known, then the value of the solution in the following time instant y_{n+1} and t_{n+1} can be estimated. The space between the points is assumed to be uniform with value

$$h = t_{n+1} - t_n \tag{8.4}$$

Fig. 8.2 Illustration of the function and its derivative at different points

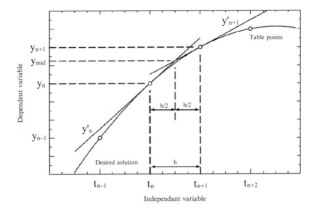

Fig. 8.3 Model solution for a control volume

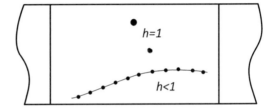

Perhaps the simplest algorithm is Euler's formula

$$y_{n+1} = y_n + hy'_n = y_n + hf(t_n, y_n) \tag{8.5}$$

which uses the future value of the solution to approximate the derivative in the n-th point. The second formula of the equation expresses the dependence of the derivative in the value, as well as the time of the function in point n. As shown in Fig. 8.2, if step size h is established small enough, the evaluated point can be accurate enough for the given purpose.

For the simulation of the front-velocity model, if the value of the time step is set at $h < 1$, several values of the concentration are calculated for a single control volume. This calculation would describe the evolution of the concentration in each control volume. However, since the substance is assumed to remain in the control volume for a single unit of time, it is not interesting to determine the evolution of the concentration in each control volume. Therefore, a step $h = 1$ is applied, and as a result, a single concentration value is obtained for each control volume. Figure 8.3 illustrates this point graphically.

In order to determine the behavior of the dynamics of the concentration during the substance's travel through the chromatographic column, an iterative process is used to calculate the value of the concentration in the current control volume, using the value obtained at the previous control volume as initial value. Figure 8.4 shows the time behavior of the concentration for the control volumes that form the

Fig. 8.4 Concentration levels in the chromatographic column. (**a**) Initial phase; (**b**) Middle phase; (**c**) Final phase

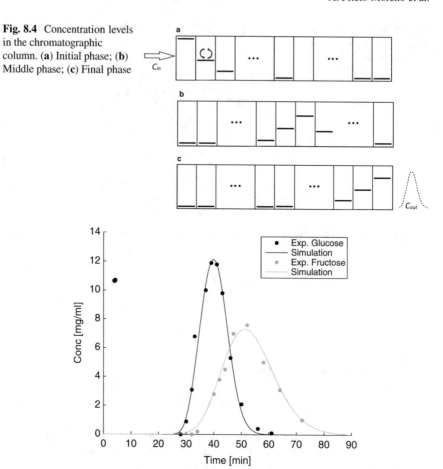

Fig. 8.5 Comparison between simulation results (*lines*) and experimental adsorption data (*points*) of glucose and fructose

entire column. In the initial stage, the mixture is injected, and as this mixture travels through the column, the adsorption process occurs; and at a final stage, the separated substance is obtained.

The simulation of the front-velocity modeling technique is compared to the experimental data available in [1]. Figure 8.5 shows a comparison between the experimental results and the data obtained by means of simulation, for which the following parameters were used: flow-rate $= 30$ ml/min, porosity $= 0.4$, injection-vol. $= 300$ ml, e injection-conc. $= 15$ mg/ml; and the Euler's method was applied to its numerical solution. With these parameters in mind, the number of control volumes to be used in the direct-problem solution is calculated as 1355.

The number of control volumes and the kinetic characteristics of substances determine the solute's residence time in the column. As can be seen in Fig. 8.5, the glucose production time is shorter because glucose has less absorption and tends to stay in the liquid phase longer. Therefore, the duration of simulation will be different for each substance, being the purpose a full profile description of the concentration of each substance.

The experimental data confirmed that a full profile description of the glucose and fructose requires a 68-min and 90-min simulation process, respectively.

8.2.3 Parameters Estimation

In order to estimate the unknown parameters of the model, the inverse problem is formulated implicitly as an optimization problem, with the intent of minimizing the squared residual function,

$$S(\vec{K}) = \left[\vec{C}_{exp} - \vec{C}_{cal}(\vec{K})\right]^T \left[\vec{C}_{exp} - \vec{C}_{cal}(\vec{K})\right] = \vec{R}^T \vec{R} \qquad (8.6)$$

where $\vec{C}_{exp}$ is the experimental solute concentration vector, $\vec{C}_{cal}$ is the calculated value vector, $\vec{K} = (k_1, k_2)^T$ is the unknown parameters vector to be determined, and the residual vector $\vec{R}$ corresponds to

$$\vec{R} = \vec{C}_{exp} - \vec{C}_{cal}(\vec{K}) \qquad (8.7)$$

The solution of the inverse problem $\vec{K}^*$ minimizes the norm given by Eq. (8.6), that is

$$\min_{\vec{K}} S(\vec{K}) = S(\vec{K}^*) \qquad (8.8)$$

The optimization method used for the solution of the inverse problem was simulated annealing (SA) [12, 20]. However, other methods may be applied, as demonstrated in [6, 14, 17].

8.2.3.1 Simulated Annealing Method

This method is based on a statistical reasoning mechanism that can be used to successfully simulate a system of atoms in equilibrium at a given temperature. In every step of this algorithm, a small random displacement of an atom takes place, and its resulting energy variation ΔE is calculated. If $\Delta E < 0$, then the displacement is accepted, and the configuration with the displaced atom is used as the starting point in the following step. If $\Delta E > 0$, the new configuration can be accepted in accordance to the Boltzmann probability

$$P(\Delta E) = \exp\left(-\frac{\Delta E}{k_B T}\right) \qquad (8.9)$$

Under the criterion defined by Metropolis in [16], a p uniformly distributed random number in the interval $[0, 1]$ is generated. This number is then compared with $P(\Delta E)$. The new configuration is accepted if $p < P(\Delta E)$; otherwise it is rejected, and the previous configuration is used again as the starting point in the following step.

Instead of simulating energy and defining configurations for a set of variables $\{P_i\}$, $i = 1, 2, \ldots, N_p$, where N_p represents the number of unknown parameters to be estimated, the Metropolis procedure uses the objective function $S(\vec{K})$ given by the Eq. (8.6) to generate a collection of configurations of an optimization problem for a temperature T that acts as control parameter. In this procedure, the system to be optimized is first "melt" at a high "temperature," followed by gradual "temperature" drops down to a system "solidification," where no more changes occur.

The main control parameters of the algorithm are: The initial "temperature" T_0, the cooling rate r_t, number of steps traveling over the elements of vector $\vec{K}$, the number of times that the procedure is repeated before the "temperature" is reduced N_s, and the number of minimal points N_t (one for each temperature), that are compared and used as the stopping criterion if they all satisfy a tolerance ϵ, N_ϵ.

Just like all the other global optimization methods, SA entails a high computational cost. Add to the above the time required for the generation of a direct-problem solution in order to obtain a system's response. In the avoidance of a major delay in parameters estimation, an iterative approach has been used herein. In this case, the algorithm is configured to execute few cycles, after which completion, the algorithm is restarted using its previous results as its initial condition. Consequently, the initial convergence speed attributed to this algorithm can be exploited. The iterative process stops when the difference between two consecutive solutions is less than a stated value or when a specific number of repetitions is reached.

8.3 Accelerated Simulation

Many proposals that combine different estimation methods have been put forward in an effort to reduce the number of calls to the objective function. However, these proposals have not addressed the possibility of reducing the computational time required by the direct-problem solution as the slow phase of the estimation process. Such a solution is the largest time consumer in any parameters estimation strategy.

8.3.1 Procedure for Selecting the Solution Step

This section discusses a procedure for the assessment of the influence exercised by the accelerated direct-problem solution on the accuracy of model parameters estimation. This procedure consists of two steps; i.e., (1) the study of the impact of any increased solution step on the solution time; and (2) the assessment of the impact by the direct-problem solution built-in error on the estimation accuracy. This step helps to establish the highest increase in the solution step without affecting the estimation.

Step 1   *Study of the direct-problem solution time*

1. Solve the direct problem by establishing different integration steps in the solution method. Repeat several times, at least 10, for each established step, and record the solution times for each execution.
2. Using statistical techniques, check for any significant decrease in solution time.
3. Determine the built-in error of the direct-problem solution as compared with experimental data.

Step 2   *Assessment of the inverse-problem solution accuracy*

1. Complete the estimation process of the model parameters by applying the integration procedure defined in Step 1 to the direct-problem solution. Repeat several times, at least 10, for each integration step.
2. Using statistical techniques, check for any significant differences among the parameters that were estimated for each integration step.

As a result of this procedure, the largest integration step that does not significantly increase the difference in the estimation may be established; and consequently, the estimation process can be accelerated without loss of accuracy.

8.3.2 Statistical Evaluation

When different procedures are tested, it is necessary to determine which has the best performance. Intuitively, the most accurate or least erring procedure can be selected; however, these average measures solely represent estimations in the real population for future data. Furthermore, the performance of the experiments may feature wide variances. Therefore, the means of the actual performance measurements for the different procedures may seem different; however, these differences may not be statistically significant. The above assertion points to the need for the application of significance statistical tests in order to determine whether or not a real difference exists among the means of the performance measures of the different procedures.

In this case, a simulation procedure that uses a step $h = 1$ for the direct model solution is established. In statistical terms, this procedure acts as the control method. The alternative method would be the simulation procedure using other values of the aforementioned step. For this situation, a test that compares the performance between two algorithms must be conducted. For the purpose of this comparison, two hypotheses are defined; i.e., the null hypothesis H_0 under which no difference exists between the procedures; in other words, the observed differences are merely random; and the alternative hypothesis H_1 that represents the presence of differences [8].

Specific methods are available to compare two procedures. The statistical test t [18], preferably its version for paired samples, and its non-parametric counterpart: the Wilcoxon Signed Ranks test [18] are the most widely used.

8.4   Results and Discussions

In the chromatographic column process that was applied to illustrate the proposed procedure and validate the hypothesis of estimation time reduction by accelerating the direct-problem simulation, two substances were considered: glucose and fructose. These substances are analyzed separately under the same procedure.

Figure 8.6 illustrates the proposed idea to increase the step in the direct-problem solution. For the front-velocity-based modeling, this idea means a decrease in the number of control volumes added to the solution. With the value $h = 2$, the concentration is calculated for every two control volumes, thus indicating that the concentration in vc_3 is calculated from the concentration value in vc_1, and so on. The same applies to steps $h = 5$ and $h = 10$, thus indicating that the concentration is calculated for each five and ten control volumes, respectively.

8.4.1   Direct Solution

This section will apply the study indicated in Step 1 of the proposed procedure. Here, the front-velocity model will be simulated for steps $h = 1, 2, 5, 10\, and\, 15$. These steps mean that $vc = 1355, 677, 271, 135\, and\, 90$ control volumes, respectively, will be applied to the direct-problem solution. For the purpose of statistically assessing the simulation time of the direct problem, the simulation was run 15 times for each step.

Fig. 8.6 Different integration steps used in the direct-problem solution

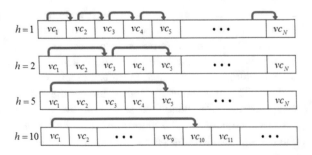

Table 8.1 CPU mean time for the glucose direct-problem solution

Step	$h = 1$	$h = 2$	$h = 5$	$h = 10$	$h = 15$
Time [ms]	3032	719	138	42	21

Table 8.2 CPU mean time for the fructose direct-problem solution.

Step	$h = 1$	$h = 2$	$h = 5$	$h = 10$	$h = 15$
Time [ms]	4344	1095	186	56	35

Tables 8.1 and 8.2 show the average time consumed by the simulation with each of the steps under study, in order to generate the complete substance concentration profile.

The procedure indicates that the differences among the simulation times could be validated using statistical tests; however, the differences among the average times are so significant that they cannot be considered the result from random phenomena.

Figures 8.7 and 8.8 show the results from the direct-problem simulation for each of the steps under study, as well as the quadratic error for the individual steps, as compared with the experimental data pertaining to both substances.

As the step considered in the direct-problem solution is increased, the accuracy of the simulation decreases in relation to the experimental data. However, for each substance, the quadratic error of simulation behaves differently. For the glucose, the error increases steadily, and as the step increases, this error grows larger. For the fructose, the influence exercised by the increased step in the direct-problem solution over the simulation error is smaller, and an increase in the step does not always imply an error increment.

8.4.2   Parameters Estimation

This section shows the application of Step 2 of the proposed procedure. Under this step, the differences of the estimated parameters are studied. For the statistical assessment, the estimation process was repeated 10 times. The estimation method was configured in a manner that a maximum of 1351 calls to the objective function were made. This number may not be reached if the alternative stopping condition of a less than 10^{-6} difference in two consecutive estimations is satisfied. Since the estimation uses an iterative process, three repetitions are set in order to obtain the final value of the estimated parameters; therefore, a maximum of 4053 calls will be made to the objective function, and each call to the objective function corresponds to a solution of the direct problem.

Tables 8.3 and 8.4 show the average values of the mass-transfer kinetic parameters obtained for each integration step, as well as the duration of the estimation process.

Fig. 8.7 Comparison
between the simulation
results and experimental data
for glucose, Eq. (8.6), for the
different number of
integration steps.
(**a**) Direct-problem solution;
(**b**) Simulation quadratic error

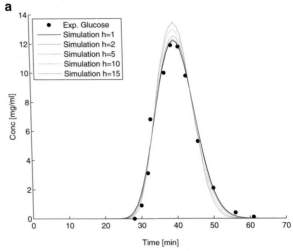

Direct-problem solution.

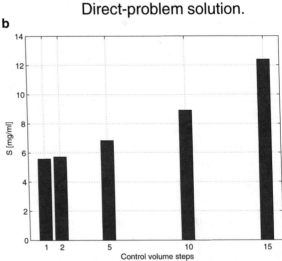

Simulation quadratic error.

In order to determine if significant differences exist among the estimated parameters, the t test poses an inconvenience: unless the sample size is large enough (~ 30 samples), this test requires that the differences among the random variables subjected to comparison be normally distributed [7]. Since the random samples subject to comparison are the result from 15 repetitions, the use of the Wilcoxon test becomes advisable. For this purpose, a null hypothesis H_0 is established as follows: the differences between the results from the control procedure, in this case, the direct-problem solution using $h = 1$, and the results from alternative procedure, in this case, other h values, are attributed to random phenomena. For the difference to be regarded as significant, the statistician's value (T) must be less than or equal

Fig. 8.8 Comparison
between simulation results
and experimental data for
fructose, Eq. (8.6) for the
different number of
integrations steps.
(**a**) Direct-problem solution;
(**b**) Simulation quadratic error

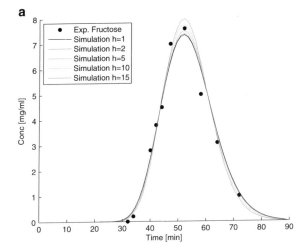

Direct-problem solution.

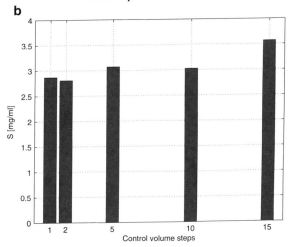

Simulation quadratic error.

Table 8.3 Parameters estimation results for the glucose

Step	$h = 1$	$h = 2$	$h = 5$	$h = 10$	$h = 15$
Time [s]	12411	3146	505	124	52
k_1 [min^{-1}]	0.01864	0.01882	0.01785	0.01668	0.01503
k_2 [min^{-1}]	0.03147	0.03162	0.02996	0.02766	0.02473

to the critical value of the test (T_{th}), for a predefined significance level. In this case,
a significance level $\alpha = 0.05$ is used. Since the test involves 10 observations, this
implies that the critical value for the test is $T_{th} = 8$.

Table 8.4 Parameters estimation results for the fructose

Step	$h = 1$	$h = 2$	$h = 5$	$h = 10$	$h = 15$
Time [s]	15647	4094	542	152	73
k_1 [min^{-1}]	0.01551	0.01577	0.01530	0.01487	0.01348
k_2 [min^{-1}]	0.01405	0.01399	0.01350	0.01311	0.01176

Table 8.5 Wilcoxon test for significant differences in the k_1 estimation for glucose

Test	h_1 vs h_2	h_1 vs h_5	h_1 vs h_{10}	h_1 vs h_{15}
Statistic T	27	11	6	0

Table 8.6 Wilcoxon test for significant differences in the k_2 estimation for glucose

Test	h_1 vs h_2	h_1 vs h_5	h_1 vs h_{10}	h_1 vs h_{15}
Statistic T	25	11	5	0

Table 8.7 Wilcoxon test for significant differences in the k_1 estimation for fructose

Test	h_1 vs h_2	h_1 vs h_5	h_1 vs h_{10}	h_1 vs h_{15}
Statistic T	27	21	15	4

Table 8.8 Wilcoxon test for significant differences in the k_2 estimation for the fructose

Test	h_1 vs h_2	h_1 vs h_5	h_1 vs h_{10}	h_1 vs h_{15}
Statistic T	27	21	15	4

Tables 8.5 and 8.6 show the values obtained for the Wilcoxon test for glucose.

As can be observed in both tables, the Wilcoxon test indicates that there are no significant differences in the estimated parameters, when a step up to $h = 5$ is used for the direct-problem solution. This evidences that the direct-problem solution can be accelerated without the built-in error affecting the parameters estimations. In other words, the optimization method applied to solve the inverse problem may continue to generate correct estimates, in spite of an increase in the built-in error of in the direct-problem solution.

Tables 8.7 and 8.8 show the same study for fructose.

Both tables suggest that it is possible to accelerate the solution with an increase in the step to $h = 10$. These results are consistent with the simulation error values shown in Fig. 8.8, where the significant increment of the error was produced for $h = 15$.

These results show that the estimation time of the mass-transfer kinetic parameters may be significantly reduced, as compared with the initially required estimation time, without any significant differences among the estimated values. These results also validate the proposal to accelerate the direct-problem solution as an strategy to cut down delays in the estimation process. The proposed procedure successfully demonstrated its capacity in determining the maximum acceleration of the solution that may be reached for each model, without affecting the accuracy of the solution.

Although the numerical results were obtained for specific conditions of the chromatography process, the proposed procedure can be applied to other parameter values.

8.5 Conclusions

This chapter discussed a procedure designed to decrease the time required by the parameters estimation process by accelerating the direct-problem solution.

The procedure helped assess the existing relationship between the direct-problem solution time and the solution step, as well as the influence exercised by the built-in error of the direct-problem solution over the accuracy of the estimated parameters.

The results from the application of the proposed procedure to the estimation of the mass-transfer kinetic parameters in a chromatographic column, as modeled using the front-velocity method, showed that the estimation time can be significantly reduced. Therefore, the proposed acceleration of the direct-problem solution as a way to cut down the estimation time is another valid approach to be considered for this purpose.

Acknowledgements The authors acknowledge the financial support provided by the Brazilian Agencies FAPERJ, Fundação Carlos Chagas Filho de Amparo à Pesquisa do Estado do Rio de Janeiro; CNPq, Conselho Nacional de Desenvolvimento Científico e Tecnológico; and CAPES, Coordenação de Aperfeiçoamento de Pessoal de Nível Superior, as well as the Ministerio de Educación Superior de Cuba (MES/Cuba).

References

1. Azevedo, D.C.S., Rodrigues, A.: SMB chromatography applied to the separation/purification of fructose from cashew apple juice. Braz. J. Chem. Eng. **17**(4–7), 507–516 (2000)
2. Bihain, A.J., Silva-Neto, A.J., Llanes-Santiago, O., Afonso, J.C., Câmara, L.D.: The front velocity modelling approach in the chromatographic column characterization of glucose and fructose separation in SMB. Trends Chromatogr. **2**, 57–73 (2012)
3. Brittes, R., França, F.H.R.: A hybrid inverse method for the thermal design of radiative heating systems. Int. J. Heat Mass Transf. **57**(1), 48–57 (2013)
4. Burden, R.L., Faires, J.D.: Numerical Analysis. Brooks Cole, Independence (2010)
5. Câmara, L.D.: Stepwise model evaluation in simulated moving-bed separation of ketamine. Chem. Eng. Technol. **37**(2), 1–10 (2014)
6. Cuco, A.P.C., Silva Neto, A.J., Campos Velho, H.F., de Sousa, F.L.: Solution of an inverse adsorption problem with an epidemic genetic algorithm and the generalized extremal optimization algorithm. Inverse Prob. Sci. Eng. **17**(3), 289–302 (2009)
7. Demsar, J.: Statistical comparisons of classifiers over multiple data sets. J. Mach. Learn. **7**, 1–30 (2006)

8. Derrac, J., García, S., Molina, D., Herrera, F.: A practical tutorial on the use of nonparametric statistical tests as a methodology for comparing evolutionary and swarm intelligence algorithms. Swarm Evol. Comput. **1**(1), 3–18 (2011)
9. Galski, R.L., de Sousa, F.L., Ramos, F.M., Silva Neto, A.J.: Application of a GEO + SA hybrid optimization algorithm to the solution of an inverse radiative transfer problem. Inverse Prob. Sci. Eng. **17**(3), 321–334 (2009)
10. Guiochon, G.: Preparative liquid chromatography. J. Chromatogr. A. **965**(1–2), 129–161 (2002)
11. Jahanshahloo, G., Soleimani-Damaneh, M., Mostafaee, A.: On the computational complexity of cost efficiency analysis models. Appl. Math. Comput. **188**(1), 638–640 (2007)
12. Laarhoven, P.J.M., Aarts, E.H.L.: Simulated Annealing: Theory and Applications. Kluwer Academic, Norwell (1987)
13. Lage, F.L., Cuco, A.P.C., Folly, F.M., Soeiro, F.J.C.P., Silva Neto, A.: Stochastic and hybrid methods for the solution of an inverse mass transfer problem. In: III European Conference on Computational Mechanics (2006)
14. Lugon, J., Silva Neto, A.J., Santana, C.C.: A hybrid approach with artificial neural networks, Levenberg-Marquardt and simulated annealing methods for the solution of gas-liquid adsorption inverse problems. Inverse Prob. Sci. Eng. **17**(1), 85–96 (2009)
15. Marquardt, D.W.: An algorithm for least-squares estimation of non-linear parameters. J. Soc. Ind. Appl. Math. **11**, 431–441 (1963)
16. Metropolis, N., Rosenbluth, A.W., Rosenbluth, M.N., Teller, A.H., Teller, E.: Equation of state calculations by fast computing machines. J. Chem. Phys. **21**, 1087–1092 (1953)
17. Pereyra, S., Lombera, G.A., Urquiza, S.A.: Modelado numérico del proceso de soldadura fsw incorporando una técnica de estimación de parámetros. Revista Internacional de Métodos Numéricos para Cálculo y Diseño en Ingeniería **30**(3), 173–177 (2014)
18. Sheskin, D.J.: Handbook of Parametric and Nonparametric Statistical Procedures. Chapman and Hall/CRC, Boca Raton (2011)
19. Silva Neto, A.J., Soeiro, F.J.C.P.: Solution of implicitly formulated inverse heat transfer problems with hybrid methods. In: Proceedings Second MIT Conference on Computational Fluid and Solid Mechanics, pp. 2369–2372 (2003)
20. Silva Neto, A.J., Becceneri, J.C.: Técnicas de Inteligência Computacional Inspiradas na Natureza: Aplicação em Problemas Inversos em Transferência Radiativa. Sociedade Brasileira de Matemática aplicada e Computacional (SBMAC), **41**, 122 (2009)

Chapter 9
Effects of Antennas on Structural Behavior of Telecommunication Towers

Patricia Martín Rodríguez, Vivian B. Elena Parnás, and Angel Emilio Castañeda Hevia

Abstract The communication lattice towers must be designed to resist wind forces and support antennas under heavy working conditions. In the past few years, a number of communication lattice towers have collapsed in Cuba as a consequence of strong winds. Antennas modify the wind flow around their tower and act as a screen against wind loads and transmit strong efforts to its members. The aim of this chapter is to assess the effect of antenna locations on some structural parameters of the tower by means of physical and numerical experiments with computational applications. A physical experiment in a wind tunnel was conducted in order to obtain the drag coefficients of a tower holding dish antennas. Finite element method models of the tower with the implementation of SAP2000 code were generated to obtain forces and displacements on the members, and natural periods of the structure. The numerical experiment consisted of a 2^3 factorial experiment, where the independent variables were horizontal position, vertical position, and number of dishes on the tower. The results show that the vertical position of dish antennas has the greatest influence on the structural behavior of their supporting tower.

Keywords Telecommunication tower • Computational design • Numerical experiment • Wind forces • Structural analysis • Antennas • Wind tunnel

9.1 Introduction

The free-standing communication lattice towers must be designed to resist wind forces and support antennas under heavy working conditions. In recent years, a number of communication lattice towers erected in Cuba have collapsed as a consequence of strong winds [4], thus causing economic and social losses for the country. A study of failures confirmed that many of the collapsed towers were supporting several antennas.

P.M. Rodríguez • V.B.E. Parnás (✉) • A.E.C. Hevia
Facultad de Ingeniería Civil, Instituto Superior Politécnico José Antonio Echeverría (CUJAE),
Marianao, La Habana, CP 19390, Cuba
e-mail: patriciamr@civil.cujae.edu.cu; vivian@civil.cujae.edu.cu; ecashevia@civil.cujae.edu.cu

© Springer International Publishing Switzerland 2016
A.J. da Silva Neto et al. (eds.), *Mathematical Modeling and Computational Intelligence in Engineering Applications*, DOI 10.1007/978-3-319-38869-4_9

The internal forces of the tower elements are modified by the shape, weight, and disposition of antenna installations. Antennas act as a screen against wind flow, thus inducing additional stress on the tower structure. The mass concentration in different parts of the structure also changes the dynamic behavior against wind flows. Towers were designed to support a small number of antennas. However, communication development has given rise to an increase in the number of antennas installed on existing structures, and the structural behavior of the supporting antennas is changing accordingly.

The effect of antenna locations on the structural behavior of free-standing lattice towers has been poorly addressed in the international literature. Most research works deal mainly with the presence of antennas in the assessments of seismic load [1, 6], but not wind action. The surveys of antenna effects on the aerodynamic behavior of lattice towers have focused on the determination of aerodynamic coefficients and interference factors, by means of tunnel tests [3, 5].

The purpose of this chapter is to assess the effect of antenna locations on some structural parameters of supporting towers by means of physical and numerical experiments with computational applications. A physical experiment was conducted to obtain drag coefficients of antenna supporting towers. These coefficients were used as input data for the calculation of the structural behavior of the tower in the numerical experiment.

9.2 Methodology

For the purpose of describing the antennas that are most frequently used in Cuban communications, data were collected from 26 existing communication lattice towers. The antennas on the towers were identified as UHF, VHF, and dishes. It was observed that the UHF and VHF antennas were similar in number and position on the tower, while dish antennas had a random distribution in terms of numbers, positions, and heights on the tower structure. Two kinds of experiments were necessary to be conducted in this study; i.e., a computational numerical experiment and a wind tunnel experiment. In order to determine the influence of the horizontal position of dishes on a cross section of their tower, drag coefficient values of antenna supporting towers were obtained through a wind tunnel experiment [7, 8], while a numerical experiment in a computational model was conducted to assess the influence of the antenna locations on some structural parameters of the towers.

9.2.1 Wind Tunnel Experiment

The communication tower selected for this study is a typical free-standing Cuban square-section design called "Najasa." The cross-section width decreases as it reaches a height of 45 m. From the 45-m height to the top (60 m), the cross section remains constant. Its members are angle steel, as shown in Fig. 9.1a. The tower was

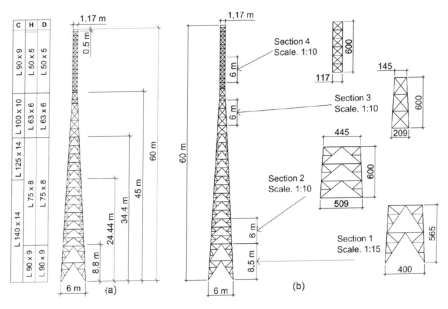

Fig. 9.1 a) Sketch of prototype and dimensions of the tower members, **b)** Sketch of tower sections studied at the wind tunnel

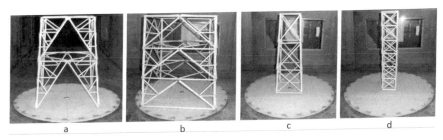

Fig. 9.2 Tower section models at wind tunnel, **(a)** section 1 **(b)** section 2 **(c)** section 3 **(d)** section 4

divided into four sections for the wind tunnel test, as shown in Figs. 9.1b and 9.2. The applied geometrical scales were 1:10 for Sects. 9.2, 9.3, and 9.4, and 1:15 for Sect. 9.1. The dish antenna chosen for this study was 2-m wide, and its detailed characteristics were obtained from an RFS catalog [10]. Experimental tests were carried out at the "Prof. Joaquim Blessman" boundary layer tunnel located at the Aerodynamic Construction Laboratory at the Federal University of Rio Grande do Sul, Brazil. The features of this tunnel can be seen in reference [2].

The data of dish antennas were collected on free-standing communication towers in Cuba and drag coefficients for the "Najasa" tower with dishes antennas were obtained from the wind tunnel test [7]. From results of wind tunnel, three study cases with single and double antennas (Case 1, Case 3, and Case 6) were selected for application to the computational model, as shown in Fig. 9.3. Drag forces were measured with a unidirectional balance, coupled with strain gages. Two full

Fig. 9.3 Selected study cases with single and double antennas for the numerical experiment (*top*) and their pictures at wind tunnel test (*bottom*)

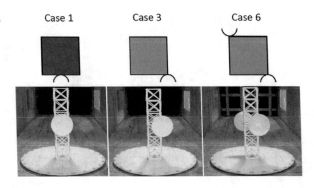

Table 9.1 Experimental drag coefficients

Case	Wind direction	Drag coefficients			
		Section 1	Section 2	Section 3	Section 4
1	0°	3.62	3.72	3.79	3.55
3	45°	3.72	3.58	3.81	3.67
6	135°	3.9	4.13	4.72	4.57

Wheatstone bridges were mounted at the balance, with four strain gages each, in order to create redundancy at the measurement, increasing the reliability of the drag forces obtained in the test. The balance resolution was 0.41 g. The measurements of drag forces were taken in a 0.2 %.

The values of the drag coefficients for each section of the tower tested at the wind tunnel are shown in Table 9.1.

9.2.2 Numerical Experiment

The numerical experiment was a 2^3 factorial design. Three independent variables were defined: horizontal position, vertical position, and number of dishes on the tower. The dependent variables were defined as axial forces on members, displacement at the top of the tower, and natural period.

Most of the towers sampled for data collection were supporting a variety of antennas, while a small number of towers had dish antennas only. In order to compare the influence of dish antennas at towers with and without other antennas (as UHF and VHF), two factorial-design experiments were conducted. Figure 9.4 shows the first experiment with a maximum number of UHF and VHF antennas (24 and 12 panels, respectively) located at the top of the tower. Figure 9.5 shows the second experiment without UHF and VHF antennas. Sixteen numerical models were processed to obtain the dependent variables.

The independent variables were set at two levels, as discussed below. For the independent variable, "number of dishes," six dishes represented the maximum

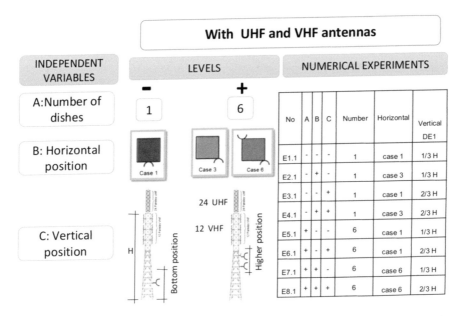

Fig. 9.4 Factorial design of experiments with UHF and VHF antennas

No	A	B	C	Number	Horizontal	Vertical DE1
E1.1	-	-	-	1	case 1	1/3 H
E2.1	-	+	-	1	case 3	1/3 H
E3.1	-	-	+	1	case 1	2/3 H
E4.1	-	+	+	1	case 3	2/3 H
E5.1	+	-	-	6	case 1	1/3 H
E6.1	+	-	+	6	case 1	2/3 H
E7.1	+	+	-	6	case 6	1/3 H
E8.1	+	+	+	6	case 6	2/3 H

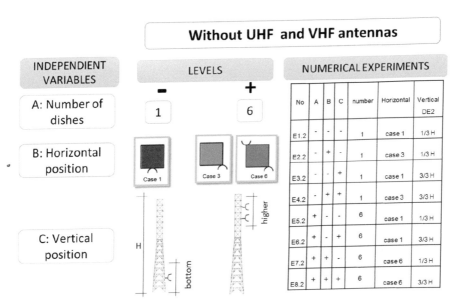

Fig. 9.5 Factorial design of experiments without UHF and VHF antennas

No	A	B	C	number	Horizontal	Vertical DE2
E1.2	-	-	-	1	case 1	1/3 H
E2.2	-	+	-	1	case 3	1/3 H
E3.2	-	-	+	1	case 1	3/3 H
E4.2	-	+	+	1	case 3	3/3 H
E5.2	+	-	-	6	case 1	1/3 H
E6.2	+	-	+	6	case 1	3/3 H
E7.2	+	+	-	6	case 6	1/3 H
E8.2	+	+	+	6	case 6	3/3 H

level and one dish represented the minimum level. For the independent variable, "horizontal position," the position featuring the highest value of drag coefficient, as defined in Cases 3 and 6 of the wind tunnel experiment, was set as maximum level and the position with the lowest drag coefficient, as defined in Case 1, was set as minimum level. For the independent variable, "vertical position," the highest possible position of any dish antenna at the tower was set as maximum level, and bottom positioned dish installation at the tower was set as minimum level, as shown in Figs. 9.4 and 9.5.

In order to identify the influence of the independent variables, and their interaction, with the values of the dependent variables, the following hypotheses were assumed:

H_o: independent variable (A, B, or C) or interaction of variables (AB, BC, or AC) has *no* influence on the value of dependent variable x.
H_1: independent variable (A, B or C) or interaction of variables (AB, BC or AC) has influence on the value of dependent variable x.

The results are based on the determination of the p-value that indicates the probability of rejection of hypothesis H_o. To identify the weight of independent variables and their interaction with the values of the dependent variables, a regression for each dependent variable was completed. The coefficients of adjustment for the regression evidenced the weight of each independent variable (B1, B2, and/or B3) and the significance of the interaction between independent variables for the values of the dependent variables (B4, B5, and/or B6). A statistical tool ("Statgraphics") was used for the regression process.

9.2.3   Structure and Wind Load Modeling

A typical "Najasa" tower was modeled as a spatial truss. The members of the truss were modeled as linear elements using FEM code SAP 2000 V14. The columns were considered continuous from top to bottom, while the horizontal and diagonal members were considered with pinned ends. The structural material was steel, with a 250 MPa yield tension and a 400 MPa ultimate stress. These properties were considered elastic and free from variation over time.

For the numerical experiment, the wind load on the tower was obtained using Cuban wind code NC: 285-2003 [9], which applies to spatial latticed structures subjected to extreme winds. However, the drag coefficients were taken from the wind tunnel tests, as shown in Table 9.1. The wind directions were 0°, 45°, and 135°. These directions generated the highest wind-load values obtained for the dish positions in Cases 1, 3, and 6.

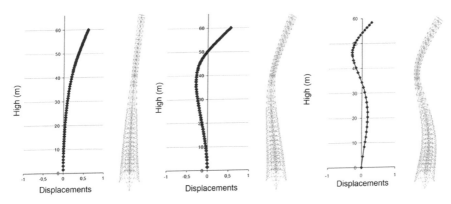

Fig. 9.6 Shape modes for a "Najasa" tower without antennas

Table 9.2 Natural periods and frequencies for tower models with UHF and VHF antennas

No	T (s)	f (Hz)
E0.1	1.251	0.799
E1.1	1.251	0.799
E2.1	1.251	0.799
E3.1	1.252	0.799
E4.1	1.252	0.798
E5.1	1.251	0.799
E6.1	1.255	0.797
E7.1	1.251	0.799
E8.1	1.254	0.797

9.2.4 Modal Analysis

A modal analysis was implemented to all the models of the numerical experiment. The purpose of this analysis was to define the dynamic characteristics of the tower and estimate the variations of natural periods due to the presence of antennas. Natural frequencies of the tower were calculated through the eigenvector method. Masses of each member of the structure were concentrated at the intersection points between columns and braced members. Masses of antennas were assigned as concentrated masses at local points of the tower.

In regard to the dynamic characteristics of the tower, the well distance between shape modes and the first mode as representative of the fundamental period of the lattice tower was confirmed, as shown in Fig. 9.6.

Tables 9.2 and 9.3 show the values of natural periods for the experiments designed with and without UHF and VHF antennas. As may be observed, the natural periods remain similar with the variation of number and position of dish antennas on the tower. It is also observed that models with UHF and VHF antennas at the top of the tower have a higher natural period than those without UHF and VHF antennas.

Table 9.3 Natural periods and frequencies for tower models without UHF and VHF antennas.

No	T (s)	f (Hz)
E0.2	0.732	1.366
E1.2	0.732	1.367
E2.2	0.732	1.367
E3.2	0.741	1.349
E4.2	0.755	1.324
E5.2	0.732	1.366
E6.2	0.786	1.272
E7.2	0.732	1.366
E8.2	0.764	1.310

9.3 Discussion of Results

Axial forces were obtained for all members of the tower that was divided into four vertical sections, consistently with the variation of the cross-section members, as shown in Fig. 9.1a. Vertical Sect. 9.1 was the nearest to the bottom of the tower. Three elements were studied: columns, braced members, and diagonals. The processed axial forces were the highest values on members (tension and compression) for each vertical section. Displacement rates were calculated for the top of the tower. The natural period of the first mode was considered to be a dependent variable.

9.3.1 Factorial Design of Experiments with UHF and VHF Antennas

Axial force at members: As shown by Table 9.4, the independent variables that influence axial forces at columns were the vertical position and the number of dishes and their related interaction at the bottom of the tower (Vertical Sects. 9.1 and 9.2). The heaviest weighing variables were the number of dishes for the bottom section and vertical position for Vertical Sect. 9.2, as shown in Table 9.5. Axial forces at diagonals and braced members did not show any 5 % significance level of influence by independent variables; hence, no weight was obtained for any variable.

Displacements: The vertical position of dishes was observed to exert a 5 % significance level of influence over the linear displacements of the tower at 45° wind direction, as shown in Table 9.6. The heaviest weighing independent variables were vertical position, for linear displacements, and number of dishes, for angular displacements, as illustrated in Table 9.7.

Natural periods: Table 9.8 shows the p-values of independent variables for natural period of the tower. Vertical position was observed as the variable with a level of significance of 5 %. Weight of independent variables and interaction showed in Table 9.9 indicates that vertical position is the most significance in this dependent variable.

Table 9.4 P-value for columns members with UHF and VHF antennas

Independent variable	Dependent variable: axial force on column							
	Zone 1		Zone 2		Zone 3		Zone 4	
	Tension	Compression	Tension	Compression	Tension	Compression	Tension	Compression
A: Number of dishes	0.04	0.01	0.03	0.03	0.4	0.37	0.67	0.85
B: Horizontal position	0.17	0.06	0.43	0.38	0.64	0.62	0.52	0.53
C: Vertical position	0.04	0.01	0.02	0.02	0.17	0.15	0.4	0.35
AB	0.25	0.1	0.42	0.4	0.47	0.46	0.46	0.43
AC	0.07	0.02	0.03	0.03	0.42	0.4	0.74	0.99
BC	0.8	0.2	0.37	0.31	0.69	0.68	0.57	0.62

Table 9.5 Weight of the independent variables and their interaction at columns with UHF and VHF antennas

Section	Force	Coefficients and errors for columns												
		Average	A:Number of dishes		B:Horizontal position		C:Vertical position		AB		AC		BC	
		B0	B1	Error	B2	Error	B3	Error	B4	Error	B5	Error	B6	Error
1	T	1141.1	40.2	4.6	−8.6	4.6	38	4.6	−5.7	4.6	22.3	4.6	−0.7	4.6
	C	−1232.1	−42.7	1.4	7.1	1.4	−36.7	1.4	4.5	1.4	−20.7	1.4	2.2	1.4
2	T	821.5	16.7	1.7	−1.1	1.7	28.4	1.7	1.1	1.7	16.4	1.7	−1.3	1.7
	C	−868.6	−17.1	1.4	1.1	1.4	−29.1	1.4	−1	1.4	−16.8	1.4	1.3	1.4

Table 9.6 P-values for displacements with UHF and VHF antennas

Independent variable	Dependent variable: displacement				
	Linear				Angular
	Direction 0	Direction 45	Direction 135	Maximum	Maximum
A: Number of dishes	0.3	0.06	0.32	0.12	0.06
B: Horizontal position	0.37	0.17	0.39	0.64	0.98
C: Vertical position	0.18	0.03	0.23	0.08	0.17
AB	0.54	0.33	0.45	0.4	0.51
AC	0.42	0.09	0.41	0.14	0.45
BC	0.35	0.2	0.43	0.83	0.3

9.3.2 Factorial Design of Experiments Without UHF and VHF Antennas

Axial forces at members: As Table 9.10 shows, for dependent variable "axial forces in columns," the only significant independent variable was vertical position of dishes. The weight of this variable for the tower's vertical sections is illustrated in Table 9.11. Axial forces at diagonals and braced members did not show any 5 % significance level of influence by independent variables; hence, no weight was obtained for any variable.

Displacements: Vertical position was observed as the independent variable with a 5 % significance level of influence over linear displacements at a 0 ° wind direction, as shown in Tables 9.12 and 9.13.

Natural period: No independent variable was observed to exert a 5 % significance level of influence; hence, no weight was obtained for any variables.

9.4 Conclusions

Numerical and physical experiments have been integrated by computational modeling to study the behavior of a typical "Najasa" lattice tower, and the effect of dish antennas under wind action. Data outputs from the physical experiment were used to compute the wind load on a dish-antenna holding tower, and the resulting wind load was inserted in the computational model.

The results from the numerical experiments show that the position of dish antennas influences the structural behavior of their supporting "Najasa" tower. In all the numerical experiments, the independent variables exercised their highest influence on dependent variables of "axial forces on columns" and "displacements." Dependent variables of "axial forces on diagonals" and "axial forces on braced members" receive no influence from the independent variables studied herein. The most influential independent variables on axial forces on columns and displacements were "vertical position" and "number of antennas"; hence, it is highly recommended

Table 9.7 Weight of independent variables and their interaction for displacements with UHF and VHF antennas

Displacement		Coefficients and errors for displacements												
		Average	A: Number of dishes		B: Horizontal position		C: Vertical position		AB		AC		BC	
		B0	B1	Error	B2	Error	B3	Error	B4	Error	B5	Error	B6	Error
Linear	Direction 45°	1.1	0.01	0	0	0	0.02	0	0	0	0.01	0	0	0
	Maximum	1.1	0.02	0.01	0	0.01	0.02	0.01	0	0.01	0.01	0.01	0	0.01
Angular	Maximum	15.47	10.6	1.94	−0.02	1.94	3.6	1.94	−0.93	1.94	1.12	1.94	−1.9	1.94

Table 9.8 P-value for natural periods with UHF and VHF antennas

Independent variables	Dependent variable: natural period
A: Number of dishes	0.07
B: Horizontal position	0.5
C: Vertical position	0.05
AB	0.5
AC	0.08
BC	0.5

that changes in vertical position and number of dish antennas be carefully studied for effective structural response to wind action.

Dish antennas exert no influence on the dynamic response of their holding tower because of their small mass; notwithstanding, they do modify the wind forces by increasing the axial forces on the tower members, due to their wind exposed area. Given their top position, UHF and VHF antennas can change the dynamic behavior of their holding tower, and their presence cause a significant increase in the value of natural periods.

Table 9.9 Weight of independent variables and interactions for natural period with UHF and VHF antenas

Natural period	Coefficients and errors												
	Average	A: Number of dishes		B: Horizontal position		C: Vertical position		AB		AC		BC	
	B0	B1	Error	B2	Error	B3	Error	B4	Error	B5	Error	B6	Error
	1.2523	0.0007	0.0002	−0.00008	0.0002	0.0001	0.0002	−0.00008	0.0002	0.0006	0.0002	−0.00008	0.0002

Table 9.10 P-value for columns without UHF and VHF antennas

Independent variable	Dependent variable: axial force on columns							
	Zone 1		Zone 2		Zone 3		Zone 4	
	Tension	Compression	Tension	Compression	Tension	Compression	Tension	Compression
A: Number of dishes	0.09	0.08	0.21	0.2	0.33	0.32	0.93	0.68
B: Horizontal position	0.42	0.36	0.68	0.61	0.68	0.62	0.75	0.59
C: Vertical position	0.04	0.04	0.06	0.05	0.07	0.06	0.08	0.15
AB	0.3	0.28	0.5	0.5	0.5	0.5	0.5	0.5
AC	0.17	0.16	0.21	0.2	0.33	0.32	0.93	0.68
BC	0.79	0.68	0.68	0.61	0.68	0.62	0.75	0.59

Table 9.11 Weight of the independent variables and interaction for columns without UHF and VHF antennas

Section	Force	Coefficients and errors for columns													
		Average	A:Number of dishes		B: Horizontal position		C: Vertical position		AB		AC		BC	
		B0	B1	Error	B2	Error	B3	Error	B4	Error	B5	Error	B6	Error
1	T	830,5	41,2	11,5	−7,4	11,5	87,9	11,5	−11,4	11,5	20,7	11,5	−2	11,5
	C	−907,9	−41,6	10,3	8,2	10,3	−88,7	10,3	10,8	10,3	−19,8	10,3	2,8	10,3
2	T	524,8	24	16,8	−4,7	16,8	88,8	16,8	−8,4	16,8	24	16,8	−4,7	16,8
	C	−560,1	−23,8	15,6	5,4	15,6	−90,2	15,6	7,8	15,6	−23,8	15,6	5,4	15,6
3	T	369,4	14,4	16,4	−4,5	16,4	74,3	16,4	−8,2	16,4	14,4	16,4	−4,5	16,4
	C	−386,5	−14,2	15,3	5,2	15,3	−75,7	15,3	7,7	15,3	−14,2	15,3	5,2	15,3
4	T	271,1	−0,8	15	−3,1	15	56,8	15	−7,5	15	−0,8	15	−3,1	15

Table 9.12 P-values for displacements without UHF and VHF antennas

Independent variable	Dependent variable: displacement				
	Linear				Angular
	Direction 0	Direction 45	Direction 135	Maximum	Maximum
A:Number of dishes	0.06	0.67	0.19	0.38	0.12
B: Horizontal position	0.14	0.9	0.36	0.81	0.45
C: Vertical position	0.03	0.3	0.13	0.17	0.31
AB	0.78	0.49	0.52	0.49	0.23
AC	0.07	0.7	0.21	0.41	0.13
BC	0.12	0.92	0.36	0.78	0.47

Table 9.13 Weight of independent variables and their interaction for displacements without UHF and VHF antennas

Coefficients and errors for displacements																
Displacement		Average	A:Number of dishes		B: Horizontal position		C: Vertical position		AB		AC		BC	
		B0	B1	Error	B2	Error	B3	Error	B4	Error	B5	Error	B6	Error
Linear	Direction 0	0,39	0,02	0	0,01	0	0,05	0	0	0	0,02	0	0,01	0

References

1. Amiri, G.G., Massah, S.R.: Seismic response of 4-legged self-supporting telecommunication towers. Int. J. Eng. Trans. B Appl. **20**(2), 107–126 (2007)
2. Blessmann, J.: The boundary layer tv-2 wind tunnel of the UFRGS. J. Wind Eng. Ind. Aerodyn. **10**(2), 231–248 (1982)
3. Carril, J.C., Isyumov, N., Brasil, R.M.L.R.F.: Experimental study of the wind forces on rectangular latticed communication towers with antennas. J. Wind Eng. Ind. Aerodyn. **91**(8), 1007–1022 (2003)
4. Elena-Parnás, V., Martín-Rodríguez, P., San-Miguel, D.: Modelación computacional y análisis cualitativo de fallas en el estudio de la vulnerabilidad de torres atirantadas de telecomunicaciones. Revista Cubana de Ingeniería **2**(1), 25–34 (2011)
5. Holmes, J., Banks, R.W., Roberts, G.: Drag and aerodynamic interference on microwave dish antennas and their supporting towers. J. Wind Eng. Ind. Aerodyn. **50**, 263–269 (1993)
6. Kherd, M.A.: Seismic analysis of lattice towers. Ph.D. Thesis, Civil Engineering and Applied Mechanics Department, McGill University, Montreal, Canada (1998)
7. Martín-Rodríguez, P.: Estudio Analítico Experimental de Torre Autosoportada con Presencia de Antenas bajo la Acción del Viento. Ph.D. Thesis, Departamento de Ingenierlía Civil, CUJAE (2014)
8. Martín-Rodríguez, P., Elena-Parnás, V., Loredo-Souza, A.M., Camaño-Schettini, E.B.: Estudio de modelo cubano de torre autosoportada en túnel de viento. In: II Congreso Internacional de Ingeniería Civil (17 Convención Científica de Ingeniería y Arquitectura). La Habana (2014)
9. NC-285:2003: Carga de viento. Método de cálculo (2003)
10. RFS: RFS Microwave Antennas: A Comprehensive Selection Guide, 2nd edn. (2013)

Chapter 10
Comparing Two-Level Preconditioners for Solving Petroleum Reservoir Simulation Problems

José R.P. Rodrigues, Paulo Goldfeld, and Luiz M. Carvalho

Abstract Domain decomposition ideas (proven suitable for parallelization) are combined with incomplete factorizations (which are standard in reservoir simulation) at subdomain level, with the ultimate goal of designing a scalable parallel preconditioner for addressing reservoir simulation problems. An ILU(k)-based two-level domain decomposition preconditioner is introduced, and its performance is compared with a two-level ILU(k)-block Jacobi preconditioner.

Keywords Oil reservoir simulation • Krylov methods • Domain decomposition • Block preconditioners • Two-level preconditioners • Coarse space • Parallel computing

10.1 Introduction

Consideration is given to the $(n \times n)$ linear system of algebraic equations

$$Ax = b \tag{10.1}$$

arising from the discretization of a PDE by a block-centered finite difference scheme with n blocks. Blocks and their respective indexes are identified, in a manner that Ω

J.R.P. Rodrigues
Petrobras/CENPES, Avenida Horácio Macedo, 950 - Ilha do Fundão,
Rio de Janeiro, RJ, CEP 21940-915, Brazil
e-mail: jrprodrigues@petrobras.com.br

P. Goldfeld
Instituto de Matemática, Universidade Federal do Rio de Janeiro, Caixa Postal 68530,
Rio de Janeiro, RJ, CEP 21941-909, Brazil
e-mail: goldfeld@ufrj.br

L.M. Carvalho (✉)
Departamento de Matemática Aplicada, Universidade do Estado do Rio de Janeiro (UERJ),
Rua São Francisco Xavier, 524, sala 6029D, Rio de Janeiro, RJ, CEP 20559-900, Brazil
e-mail: luizmc@ime.uerj.br

© Springer International Publishing Switzerland 2016   141
A.J. da Silva Neto et al. (eds.), *Mathematical Modeling and Computational Intelligence in Engineering Applications*, DOI 10.1007/978-3-319-38869-4_10

Fig. 10.1 2D domain partitioned into four subdomains; discretization based on a 5-point stencil

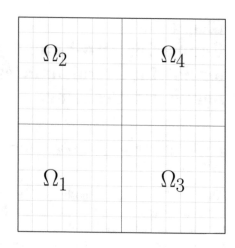

will denote either the domain of the PDE or the set of indexes $\{1, 2, \ldots, n\}$. An Ω disjoint partition is introduced; i.e.,

$$\{\Omega_J\}_{1 \leq J \leq P} \quad \text{s.t.} \quad \bigcup_{J=1}^{P} \Omega_J = \Omega \quad \text{and} \quad \Omega_I \cap \Omega_J = \emptyset \ \forall I \neq J \tag{10.2}$$

Figure 10.1 illustrates an Ω domain decomposed into subdomains Ω_J.

A local interface Γ_J associated with each subdomain Ω_J is defined as follows:

$$\Gamma_J = \{j \in \Omega_J \mid (\exists K > J \text{ and } \exists k \in \Omega_K) \text{ s.t. } (a_{jk} \neq 0 \text{ or } a_{kj} \neq 0)\} \tag{10.3}$$

where a_{ij} is the entry at the i-th row and j-th column of A. The interior subdomain is defined as

$$\Omega_J^{\text{Int}} = \Omega_J \setminus \Gamma_J \tag{10.4}$$

The (global) interface is then defined as

$$\Gamma = \bigcup_{J=1}^{P} \Gamma_J \tag{10.5}$$

Note that $\{\Gamma_J\}_{1 \leq J \leq P}$ form a disjoint partition of Γ. See Fig. 10.2.

Set $\overline{\Gamma}_J$ is defined as

$$\overline{\Gamma}_J = \Gamma_J \cup \{k \in \Gamma \mid \exists j \in \Omega_J \text{ s.t. } (a_{jk} \neq 0 \text{ or } a_{kj} \neq 0)\} \tag{10.6}$$

Notice that $\Gamma_J \subset \overline{\Gamma}_J \subset \Gamma$. $\overline{\Gamma}_J$ is the result of augmenting Γ_J with the blocks of the interface Γ whose corresponding equations/variables are connected to Ω_J in the A graph. $\overline{\Gamma}_J$ is an extended interface, see Fig. 10.3. Extended subdomains $\overline{\Omega}_J = \Omega_J^{\text{Int}} \bigcup \overline{\Gamma}_J$, are also defined; see Fig. 10.4.

Fig. 10.2 Local interfaces and interiors; the interfaces have no overlapping

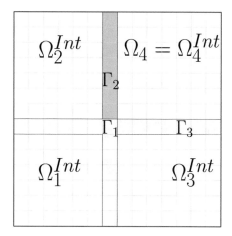

Fig. 10.3 Extended interface $\overline{\Gamma}_3$ has points in common with Γ_1

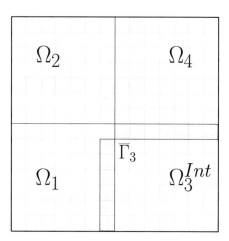

Fig. 10.4 Extended subdomain $\overline{\Omega}_2$ incorporates points from Ω_1

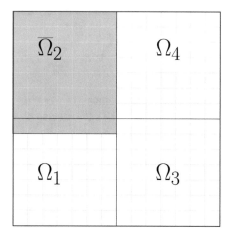

Notice that $a_{jk} = 0$ for any $j \in \Omega_J^{\text{Int}}$ and $k \in \Omega_K^{\text{Int}}$ with $J \neq K$; hence, if the equations/variables corresponding to $\Omega_1^{\text{Int}}, \ldots, \Omega_P^{\text{Int}}$ are numbered first, followed by the ones corresponding to Γ, then A will have this structure:

$$A = \begin{bmatrix} A_{11} & & & A_{1\Gamma} \\ & \ddots & & \vdots \\ & & A_{PP} & A_{P\Gamma} \\ \hline A_{\Gamma 1} & \cdots & A_{\Gamma P} & A_{\Gamma\Gamma} \end{bmatrix} \tag{10.7}$$

10.2 Methods

In this section, a two-level preconditioner that combines ideas from domain decomposition and incomplete factorization is proposed. Section 10.2.1 illustrates its fine component M_F. Finally, Sect. 10.2.2 discusses a coarse component M_C.

10.2.1 ILU(k)-Based Domain Decomposition

The fine component of the domain decomposition preconditioner described herein is based on the following block LU factorization of A

$$A = LU = \begin{bmatrix} L_1 & & & \\ & \ddots & & \\ & & L_P & \\ \hline B_1 & \cdots & B_P & I \end{bmatrix} \begin{bmatrix} U_1 & & & C_1 \\ & \ddots & & \vdots \\ & & U_P & C_P \\ \hline & & & S \end{bmatrix} \tag{10.8}$$

where $A_{JJ} = L_J U_J$ is the LU factorization of A_{JJ}, $B_J = A_{\Gamma J} U_J^{-1}$, $C_J = L_J^{-1} A_{J\Gamma}$ and

$$S = A_{\Gamma\Gamma} - \sum_{J=1}^{P} A_{\Gamma J} A_{JJ}^{-1} A_{J\Gamma} = A_{\Gamma\Gamma} - \sum_{J=1}^{P} B_J C_J \tag{10.9}$$

is the Schur complement of A with respect to the interior points. From the decomposition in (10.8), the inverse of A is shown as

$$A^{-1} = \begin{bmatrix} U_1^{-1} & & & \\ & \ddots & & \\ & & U_P^{-1} & \\ \hline & & & I \end{bmatrix} \begin{bmatrix} I & & & C_1 \\ & \ddots & & \vdots \\ & & & C_P \\ \hline & & & -I \end{bmatrix} \begin{bmatrix} I & & \\ & \ddots & \\ \hline & & S^{-1} \end{bmatrix} \begin{bmatrix} I & & \\ & \ddots & \\ \hline B_1 & \cdots & B_P & -I \end{bmatrix} \begin{bmatrix} L_1^{-1} & & & \\ & \ddots & & \\ & & L_P^{-1} & \\ \hline & & & I \end{bmatrix} \tag{10.10}$$

This chapter intends to define an M_F preconditioner that approximates the action of A^{-1} on a vector. For this purpose, suitable approximations are required for the actions of L_J^{-1}, U_J^{-1}, B_J, and C_J, $J = 1, \ldots, P$, and for that of S^{-1}, which are identified as $\tilde{L}_J^{-1}$, and S_F^{-1}, respectively. Then, the fine preconditioner is defined as:

$$
M_F = \begin{bmatrix} \tilde{U}_1^{-1} & & \\ & \ddots & \\ & & \tilde{U}_P^{-1} \\ \hline & & I \end{bmatrix} \begin{bmatrix} I & \tilde{C}_1 \\ & \vdots \\ & \tilde{C}_P \\ \hline & -I \end{bmatrix} \begin{bmatrix} I & \\ \hline & S_F^{-1} \end{bmatrix} \begin{bmatrix} I & \\ \hline \tilde{B}_1 \cdots \tilde{B}_P & -I \end{bmatrix} \begin{bmatrix} \tilde{L}_1^{-1} & & \\ & \ddots & \\ & & \tilde{L}_P^{-1} \\ \hline & & I \end{bmatrix}
$$

$$(10.11)$$

The rest of this section accurately describes how these approximations are taken.

First, $\tilde{L}_J$ and $\tilde{U}_J$ are defined as the result of the incomplete LU factorization of A_{JJ} with level of fill k_{Int}, $[\tilde{L}_J, \tilde{U}_J] = \text{ILU}(A_{JJ}, k_{\text{Int}})$. Even though $\tilde{L}_J$ and $\tilde{U}_J$ are sparse, $\tilde{L}_J^{-1} A_{J\Gamma}$ and $\tilde{U}_L^{-T} A_{\Gamma J}^T$ (which would approximate C_J and B_J^T) are not. To ensure sparsity, $\tilde{C}_J \approx \tilde{L}_J^{-1} A_{J\Gamma}$ is defined as the result of an incomplete triangular solve, by extending the definition of *level of fill* as follows.

Let v_l and w_l be the sparse vectors corresponding to the l-th columns of $A_{J\Gamma}$ and $\tilde{L}_J^{-1} A_{J\Gamma}$, respectively. Based on the solution of a triangular system by forward substitution, the components of v_l and w_l are related by

$$
w_{l_k} = v_{l_k} - \sum_{i=1}^{k-1} \tilde{L}_{J_{ki}} w_{l_i} \tag{10.12}
$$

The fill-in level of component k of w_l is defined recursively as

$$
\text{Lev}(w_{l_k}) = \min \left\{ \text{Lev}(v_{l_k}), \min_{1 \le i \le (k-1)} \left\{ \text{Lev}(\tilde{L}_{J_{ki}}) + \text{Lev}(w_{l_i}) + 1 \right\} \right\}, \tag{10.13}
$$

where $\text{Lev}(v_{l_k}) = 0$ when $v_{l_k} \ne 0$ and $\text{Lev}(v_{l_k}) = \infty$ otherwise, and $\text{Lev}(\tilde{L}_{J_{ki}})$ is the entry fill level ki in the $\text{ILU}(k_{\text{Int}})$ decomposition of A_{JJ} when $\tilde{L}_{J_{ki}} \ne 0$ and $\text{Lev}(\tilde{L}_{J_{ki}}) = \infty$ otherwise.

The approximation $\tilde{C}_J$ to $C_J = L_J^{-1} A_{J\Gamma}$ is then obtained by the designated *incomplete forward substitution* with level of fill k_{Bord}, in which any terms with fill level greater than k_{Bord} are dropped during the forward substitution process. Note that $\tilde{C}_J = \text{IFS}(\tilde{L}_J, A_{J\Gamma}, k_{\text{Bord}})$. Also note that when $k_{\text{Bord}} = k_{\text{Int}}$, $\tilde{C}_J$ results from a standard partial incomplete factorization of A. The approximation $\tilde{B}_J$ to $B_J = A_{\Gamma J} \tilde{U}_J^{-1}$ is defined analogously.

Similarly, in order to define an $\tilde{S}$ to S approximation, $F_J = \tilde{B}_J \tilde{C}_J$ is first defined followed by the determination of a fill level for the F_J entries,

$$
\text{Lev}(F_{J_{kl}}) = \min \left\{ \text{Lev}(A_{\Gamma\Gamma_{kl}}), \min_{1 \le i \le m} \left\{ \text{Lev}(\tilde{B}_{J_{ki}}) + \text{Lev}(\tilde{C}_{J_{il}}) + 1 \right\} \right\} \tag{10.14}
$$

where $m = \#\Omega_J^{\text{Int}}$ is the number of columns in $\tilde{B}_J$ and rows in $\tilde{C}_J$, $\text{Lev}(A_{\Gamma\Gamma_{kl}}) = 0$ when $A_{\Gamma\Gamma_{kl}} \neq 0$ and $\text{Lev}(A_{\Gamma\Gamma_{kl}}) = \infty$, otherwise, and $\text{Lev}(\tilde{C}_{J_{ki}})$ is the fill level according to definition (10.13) when $\tilde{C}_{J_{ki}} \neq 0$ and $\text{Lev}(C_{J_{ki}}) = \infty$, otherwise $(\text{Lev}(\tilde{B}_{J_{il}})$ is defined analogously). Next, $\tilde{F}_J$ is defined as the matrix obtained in which only the entries in F_J with level less than or equal to k_{Prod} are retained according to (10.14). This is referred to s *incomplete product* $\tilde{F}_J = \text{IP}(\tilde{B}_J, \tilde{C}_J, k_{\text{Prod}})$.

$\tilde{S}$ is then defined as

$$\tilde{S} = A_{\Gamma\Gamma} - \sum_{J=1}^{P} \tilde{F}_J \tag{10.15}$$

It should be recalled that while $\tilde{S}$ approximates S, an approximation for S^{-1} is required. Since $\tilde{S}$ is defined on the global interface Γ, it is not practical to perform ILU on it. Instead, the approach employed in [3] is applied, and a local version of $\tilde{S}$ is defined for each subdomain,

$$\tilde{S}_J = R_J \tilde{S} R_J^T \tag{10.16}$$

where $R_J : \Gamma \to \overline{\Gamma}_J$ is a restriction operator such that $\tilde{S}_J$ is the result of pruning $\tilde{S}$ in a manner that only the rows and columns associated with $\overline{\Gamma}_J$ remain. More precisely, if $\{i_1, i_2, \ldots, i_{n_{\overline{\Gamma}_J}}\}$ is a list of the nodes in Γ that belong to $\overline{\Gamma}_J$, then the k-th row of R_J is $e_{i_k}^T$, the i_k-th row of the $n_\Gamma \times n_\Gamma$ identity matrix,

$$R_J = \begin{bmatrix} e_{i_1}^T \\ \vdots \\ e_{i_{n_\Gamma}}^T \end{bmatrix}$$

Finally, an approximation S_F^{-1} to S^{-1} is defined as

$$S_F^{-1} = \sum_{J=1}^{P} T_J (L_{\tilde{S}_J} U_{\tilde{S}_J})^{-1} R_J \approx \sum_{J=1}^{P} T_J \tilde{S}_J^{-1} R_J \tag{10.17}$$

where $L_{\tilde{S}_J}$ and $U_{\tilde{S}_J}$ are given by $ILU(k_\Gamma)$ of $\tilde{S}_J$.

Here $T_J : \overline{\Gamma}_J \to \Gamma$ is an extension operator that takes values from a vector that lies in $\overline{\Gamma}_J$, scales them by $w_1^J, \ldots, w_{n_{\overline{\Gamma}_J}}^J$ (designates as weights), and places them in the corresponding position of a vector that lies in Γ. Therefore, using the aforementioned notation, the k-th column of T_J is $w_k^J e_{i_k}$,

$$T_J = \begin{bmatrix} w_1^J e_{i_1} & \cdots & w_{n_\Gamma}^J e_{i_{n_\Gamma}} \end{bmatrix}$$

Three different choices for the weights are considered, giving rise to three options for T_J, designated as T_J^{ones}, T_J^{was}, and T_J^{ras}.

The T_J^{ones} corresponds to choice $w_i^J = 1$, $i = 1, \ldots, n_{\Gamma_J}$, so that $T_J^{\text{ones}} = R_J^T$. A *counting function* is defined on the interface as

$$\mu_\Gamma = \sum_{J=1}^{P} T_J^{\text{ones}} 1_J$$

where 1_J is a vector that lies in $\overline{\Gamma}_J$ with all entries equal to 1. The i-th entry of μ_Γ, denoted $\mu_{\Gamma i}$, counts how many extended interfaces the i-th interface node belongs to.

T_J^{was} corresponds to the choice $w_i^J = \mu_{\Gamma i}^{-1}$. Note that $T_J^{\text{was}}, J = 1, \ldots, P$, form a *partition of unity*, i.e., $\sum_{J=1}^{P} T_J^{\text{was}} 1_J = 1_\Gamma$, where 1_Γ is an interface vector with all entries equal to 1.

Finally, in order to define T_J^{ras}, the following formula is established:

$$w_k^J = \begin{cases} 1, & \text{if the } i_k\text{-th node of } \Gamma \text{ belongs to } \Gamma_J \\ 0, & \text{if the } i_k\text{-th node of } Gamma \text{ belongs to } \overline{\Gamma}_J \setminus \Gamma_J \end{cases}$$

Note that $T_J^{\text{ras}}, J = 1, \ldots, P$, also form a partition of unity. The notations "RAS" and "WAS" are motivated by the Restricted and Weighted Additive Schwarz methods, see [2].

These three different choices for T_J yield three different versions of the preconditioner, which are named ONES, RAS, and WAS.

10.2.2 Coarse Space Correction

A coarse space is defined by the columns of an $(n \times P)$ matrix called R_0^T. The J-th column of R_0^T is associated with the extended subdomain $\overline{\Omega}_J$ and its i-th entry is

$$(R_0^T)_{iJ} = \begin{cases} 0, \text{if node } i \text{ is not in } \overline{\Omega}_J \text{ and} \\ \mu_{\Omega i}^{-1}, \text{if node } i \text{ is in } \overline{\Omega}_J \end{cases}$$

where the i-th entry of μ_Ω, defined as $\mu_{\Omega i}$, counts how many extended subdomains the i-th node belongs to. Notice that $(R_0^T)_{iJ} = 1 \quad \forall i \in \Omega_J^{\text{Int}}$ and the columns of R_0^T form a partition of unity, that is, their sum is a vector with all entries equal to 1.

M_C is defined using the following formula:

$$M_C = R_0^T (R_0 A R_0^T)^{-1} R_0. \tag{10.18}$$

Notice that this definition ensures that $M_C A$ is a projection onto range(R_0^T) and for symmetric positive definite A this projection is A-orthogonal. Since $R_0 A R_0^T$ is small ($P \times P$), exact LU (rather than ILU)is used when applying its inverse.

Finally, the complete preconditioner has two components: one related to the complete grid, called M_F in Eq. (10.11), and the other related to the coarse space (10.18). It is possible to subtract a third term that improves the performance of the whole preconditioner, but its computational cost increases. The combined preconditioners are written as

$$M = M_F + M_C \quad \text{or} \tag{10.19}$$

$$M = M_F + M_C - M_F A M_C \tag{10.20}$$

This formulation implies that the preconditioners will be applied additively (10.19) or multiplicatively (10.20) and may be interpreted as having two levels, see [3]. In the following sections, these preconditioners are called algebraic ILU(k)-based two-level domain decomposition, or simply ischur (for "incomplete Schur").

10.3   Analysis and Discussion

For the purpose of assessing the performance of the proposed preconditioners, an extensive set of experiments was conducted using matrices and right-hand sides arising from the discretization of PDEs modeling multiphase flow in porous media. The main characteristics of the tests are described in Sect. 10.3.1. The tests are presented in Sect. 10.3.2. Finally, the main results are discussed in Sect. 10.3.3.

10.3.1   Tests Description

The linear systems addressed herein are solved using right-preconditioned GMRES, with a restart of 30 and a zero initial guess. The stopping criterion is the relative reduction of the residual norm to 10^{-4}. The maximum number of iterations was limited to 500.

The code was written in C, using the library PETSc 3.5.4 [1], PT-Scotch 6.0.0 as the partitioner, and OpenMPI 1.6.5. The tests were performed on a parallel multicore machine with two Intel Xeon E5-2650 processors (totaling 16 cores at 2.60GHz each), 64GB of RAM, and running Debian GNU/Linux. The code was compiled with GNU GCC version 4.9.2.

Table 10.1 illustrates the number of nonzero entries in each of the six matrices that were used in the experiments. All the matrices are obtained from an actual reservoir simulator (each one is the matrix for one-Newton iteration in one timestep of a complete simulation). Matrix SPE10 arises from the simulation of the benchmark case [4], while the others involve real field models.

The block Jacobi preconditioner, described in [5], is known as "bjacobi" and the multiplicative two-level Schur preconditioner, see (10.20), as "ischur." For the sake

Table 10.1 Description of matrices

	Matrices					
	F8	U23	SPE10	U10	F11	ABL
NNZ	7.3e5	7.3e5	4.4e6	2.0e5	4.2e4	8.6e5

Table 10.2 Iterations number for 2-level ischur

	Processes						
Mat	1	2	4	8	16	32	64
F8	146	145	136	114	111	116	100
U23	71	84	83	72	63	63	66
SPE10	24	24	24	24	26	27	34
U10	52	56	56	85	135	133	154
ABL	112	119	108	105	111	147	138
F11	97	120	114	103	68	59	67

Table 10.3 Iterations ratio for 2-level bjacobi

	Processes						
Mat	1	2	4	8	16	32	64
F8	146	177	143	125	145	118	105
U23	71	98	86	79	88	76	100
SPE10	24	26	27	30	27	49	37
U10	52	58	76	111	212	170	158
ABL	112	118	108	113	135	164	174
F11	97	120	107	119	70	69	69

of comparison, the bjacobi is used with the same coarse correction as the ischur; in other words, the block Jacobi is applied as the fine part, and with the same coarse space correction that is used for ischur. Here, the RAS version of ischur is used.

10.3.2   Experiments

Two sets of experiments are presented. The first set is a comparison between the number of iterations of ischur and bjacobi. The second set shows the impact of the coarse space correction.

10.3.2.1   Comparison of Preconditioners

Table 10.2 lists the number of iterations for the ischur preconditioner and Table 10.3 lists the number of iterations for bjacobi. Table 10.4 shows the ratio between the figures for ischur and for bjacobi. Note that, when running in a single process, there are no interfaces, $\Gamma = \emptyset$. In this case, bjacobi and ischur are one and the same preconditioner, which explains why the first columns in Tables 10.2 and 10.3 are identical.

Table 10.4 Iterations ratio ischur/bjacobi, two-level preconditioners

	Processes						
Mat	1	2	4	8	16	32	64
F8	1.00	0.82	0.95	0.91	0.77	0.98	0.95
U23	1.00	0.94	0.97	0.91	0.72	0.83	0.66
SPE10	1.00	0.92	0.89	0.80	0.96	0.55	0.92
U10	1.00	0.97	0.74	0.76	0.64	0.78	0.98
ABL	1.00	1.00	1.00	0.93	0.82	0.90	0.79
F11	1.00	1.00	1.07	0.87	0.97	0.86	0.97

Table 10.5 Iterations for ischur with only the fine component

	Processes						
Mat	1	2	4	8	16	32	64
F8	182	209	221	224	223	298	NC
U23	102	106	105	107	113	147	149
SPE10	24	27	29	28	34	30	33
U10	67	74	112	243	234	NC	NC
ABL	104	107	111	132	160	197	253
F11	75	77	91	88	94	116	135

Table 10.6 Iterations ratio two/one level ischur preconditioner

	Processes						
Mat	1	2	4	8	16	32	64
F8	0.80	0.69	0.62	0.51	0.50	0.39	NC
U23	0.70	0.79	0.79	0.67	0.56	0.43	0.44
SPE10	1.00	0.89	0.83	0.86	0.76	0.90	1.03
U10	0.78	0.75	0.50	0.35	0.58	NC	NC
ABL	1.07	1.11	0.97	0.80	0.69	0.75	0.54
F11	1.29	1.56	1.25	1.17	0.72	0.51	0.50

10.3.2.2 Coarse Space Correction Impact

Table 10.5 lists the number of iterations without the use of the coarse space correction, and Table 10.6 shows the ratio between the number of iterations, with and without coarse space correction. These experiments involve ischur; however, a similar behavior is found when bjacobi is used as the fine component. In these tables NC means "no convergence," i.e., that the number of iterations of GMRES reached the limit of 500.

10.3.3 Discussion

As can be inferred from Table 10.4, for the vast majority of experiments, the Schur preconditioner needed fewer iterations than the block Jacobi preconditioner. In favor of the block Jacobi alternative, it can be argued that there is less communication between the processes during its application in a parallel architecture. This advantage may be realized by overlapping communication and computation, as reported in the specialized literature; see [3]. Another strong point of bjacobi is its simpler setup phase, as compared with the ischur construction; which means that the reduction in the number of iterations by ischur has to be important enough to compensate this difference. This subtle issue heavily depends on the underlying parallel architecture. For the time being, no conclusions can be drawn.

The coarse space correction plays an essential role in decreasing the number of iterations when the number of parallel processes increases for matrices F8, U23, and F11. Nonetheless, apparently due to the simplicity of the actual coarse space, the number of iterations does not stabilize for some experiments, as illustrated in Table 10.2 for the matrices SPE10, U10, and ABL. In an effort to address this behavior, the implementation of other coarse space corrections is required in order to incorporate, in their construction, some relevant information about the related operator; for instance, some approximate spectral data.

10.4   Future Work

New coarse spaces need to be implemented and, the complete parallel performance of ischur will be assessed. The former is related to a discussion that has common ground with multigrid problems; that is, an issue that the domain decomposition community has been dealing with for quite a few years. The latter must be based on the existing hybrid parallel architectures that combine multiprocessor machines with accelerators; such as GPUs or Xeon Phis, which means that the intensive numerical computational kernels have to be treated with care in order to achieve the necessary performance. Another important issue is the possible redesign of parts of the proposed preconditioners for adaptation to these new environments.

Acknowledgements   The third author thanks the financial support of FAPERJ, through the APQ5 2014/01 Program, E-26/111.007/2014.

References

1. Balay, S., Brown, J., Buschelman, K., Gropp, W., Kaushik, D., Knepley, M., McInnes, L., Smith, B., Zhang, H.: PETSc Web page (2013). http://www.mcs.anl.gov/petsc/
2. Cai, X.C., Sarkis, M.: A restricted additive Schwarz preconditioner for general sparse linear systems. SIAM J. Sci. Comput. **21**(2), 792–797 (1999)
3. Carvalho, L., Giraud, L., Le Tallec, P.: Algebraic two-level preconditioners for the Schur complement method. SIAM J. Sci. Comput. **22**(6), 1987–2005 (2001)
4. Christie, M., Blunt, M., et al.: Tenth SPE comparative solution project: a comparison of upscaling techniques. In: The Society of Petroleum Engineers (ed.) SPE Reservoir Simulation Symposium (2001)
5. Golub, G., van Loan, C.: Matrix Computations, 4th edn. Johns Hopkins University Press, Baltimore (2013)

Chapter 11
Assessment of the Reliability of Electrical Power Systems

Yorlandys Salgado Duarte and Alfredo M. del Castillo Serpa

Abstract This chapter assesses the reliability of a generation system (HL-I) using the Monte Carlo Non-Sequential (MCNS) simulation method and probabilistic indicators, as well as the static sense of a secured generation–transmission system (HL-II) on the basis of its voltage stability. For the purpose of mathematically modeling the system's adequacy, two-state generator models are applied; and for the purpose of the load curve, a relative cumulative frequency histogram, with polynomial interpolation among classes, is used. The security assessment discussed herein is based on the continued load flow charts obtained using the Newton–Raphson method that can monitor the voltage system's behavior to load increases. A modal analysis is also conducted in order to identify, from an overall perspective, the nodes that most contribute to loss of voltage stability in the system. Both assessments take into account the performance of the system stability limits. These limits are established on the basis of certain state variables, such as voltage range at each node, thermal limits on the transmission lines, and the reactive power limits of the generating units.

Keywords Reliability evaluation • Electrical power system • Monte Carlo simulation • Voltage stability • Probabilistic indicators • Newton–Raphson method • Modal analysis

11.1 Introduction

Modern society has a significantly high dependence on the availability of electrical supply. The basic objective of an electrical power system (EPS) is to provide energy to its customers in the most cost-effective way possible, and with a certain degree of quality and continuity [4]. It is technically and financially impossible to design an EPS that is 100 % reliable. Therefore, from their planning, engineering and operational perspectives, EPSs are designed to feature acceptable reliability levels, consistently with existing financial constraints. In order to resolve the conflict

Y.S. Duarte (✉) • A.M. del Castillo Serpa
CEMAT, Instituto Superior Politécnico José Antonio Echeverrịa (CUJAE),
Marianao, La Habana, CP 19390, Cuba
e-mail: cmtysalgado@gmail.com; acastillo47@gmail.com

© Springer International Publishing Switzerland 2016   153
A.J. da Silva Neto et al. (eds.), *Mathematical Modeling and Computational Intelligence in Engineering Applications*, DOI 10.1007/978-3-319-38869-4_11

between financial and reliability limitations, a wide range of methods and criteria have been developed for implementation in the design, planning, and operation phases of EPS's [1].

As critical components of any EPS, generators and transmission lines produce and carry large chunks of energy to consumption centers within the operational restrictions imposed on transfers, voltage and frequency; and in this respect, reliability assessment plays a relevant role. The models and methods used for reliability analysis are tailored to the attributes being studied; i.e., safety and/or adequacy [4]. Usually, distinctions are drawn between these two attribute-based studies, as discussed in references [9, 11], however, this paper combines these two attributes into a single study for more consistent results.

In this chapter, the EPS adequacy is estimated as the capacity of the generation system (HL-I) to satisfy its customer's demand for energy at all times, with regard to both scheduled downtime and reasonably expected unscheduled failures of the generation units [4]. The loss of load expectation (LOLE) and the expected energy not supplied (EENS) probabilistic indicators, as used herein, can evaluate the above-mentioned factors and quantify their risk.

This study evaluates the EPS security from the perspective of the generation–transmission (HL-II) system's response to any given contingency; in this case, the loss of system's major elements, such as transmission lines and generation units. In other words, the voltage stability in the face of certain contingencies is assessed by simulating the continued power flows in the EPS and by conducting modal analyses. Similar procedures are applied in reference [4].

11.2   Materials and Methods

11.2.1   Models and Methods for the Adequacy Assessment

11.2.1.1   Generating Unit Model

Over the past two decades, considerable works have been produced on the application of the Monte Carlo simulation to EPS reliability assessment [7]. These simulations use random-number generators [15] and probabilistic approaches to model the behavior of an EPS. The Monte Carlo simulation method for EPS reliability assessment has two classifications; i.e., Monte Carlo non-sequential (MCNS) simulation and Monte Carlo sequential (MCS) simulation [4]. This study applies the first-mentioned method. The basic principle of the MCNS simulation procedure, as applied to an EPS reliability assessment, is defined as a state sampling procedure [6], that is briefly described below.

In this approach, the state of the system is obtained by sampling the state of all its components, regardless any chronology of events. A basic sampling procedure is performed, assuming that the behavior of each component (HL-I generators) can be categorized by an even distribution in [0, 1]. Each component is represented by a two-state model; that is, each component has only two possible states: available or unavailable. Here, it is also assumed that each component's failure constitutes an independent event. The state of a system contains n components that can be expressed by the S vector, where $S = (S_1, \ldots, S_i, \ldots, S_n)$, and S_i is the state of the generation capacity of the ith component. Therefore, from the S vector, the system's state of capacity may be determined. If and when S is not equal to the total installed capacity, then the system is in a contingency state, due to a component failure.

The following steps describe the process of this method to simulate the system's installed capacity.

Step 1: A uniform random number U_i for component i is generated.
Step 2: The state of component i is determined by applying the following function:

$$S_i = \begin{cases} X \text{ if } 0 < U_i < FOR_i \\ \\ 0 \text{ if } U_i \geq FOR_i \end{cases} \tag{11.1}$$

where FOR_i is the forced outage rate [4] of component i, and X is the generating capacity of unit i.
Step 3: The system status S is obtained by applying Step 2 to every component.
Step 4: The system state is determined. If S is equal to the system's total capacity, then the system is in normal state. Otherwise, the system is in a contingency state.

11.2.1.2   Demand Curve Model

The load curve is a factor that significantly affects the reliability assessment of an EPS [12]; therefore, its simulation must be rigorous and as close to reality as possible. The load curve simulation shown herein is quite specific, as well as consistent with its application in the MCNS simulation method. From the load duration curve (LDC) [4], and with the aid of descriptive statistics, a cumulative frequency table is built, which includes values that are divided into non-overlapping classes. In this fashion, a set of n data can be summarized.

For continuous data, the classes constitute value ranges called "class intervals." The classes are selected in a manner that any data can be totally classified; that is, each data belongs in one and only one class. The definition and quantity of the classes varies from case to case. For continuous data, it is common to select the classes with equal amplitude h, as determined using the following formula:

$$h = \frac{X_{\max} - X_{\min}}{k} \tag{11.2}$$

where $X_{\max}$ is the highest value of the data and $X_{\min}$ the smallest value of the data. Each class is determined by a lower limit L_{Ii} and an upper limit L_{Si}, as follows: the i class is the $[L_{Ii}, L_{Si})$ interval, where $L_{Ii} = X_{\min} + (i-1) \cdot h$ and $L_{Si} = X_{\min} + i \cdot h$. The last class is a closed interval at both ends. The relative cumulative frequency of the i class can be calculated as follows:

$$F_{ri} = \sum_{k=1}^{i} F_k \tag{11.3}$$

From the relative cumulative frequency, the probability associated with each class can be determined; therefore, for the case of a load curve, the associated probable occurrence of a certain level of demand (D_n) may be established. It is thus possible to simulate the behavior of the curve by generating random numbers U_k that are evenly distributed between [0, 1] as follows:

If $0 \geq U_k < P_1$, then the level of demand is D_1.
If $P_1 \geq U_k < P_2$, then the level of demand is D_2.
If $P_{i-1} \geq U_k < P_i$, then the level of demand is D_i.
If $P_{i=n-1} \geq U_k \geq 1$, then the level of demand is D_n.

This form of simulating the load curve poses a limitation, as established above: the level of demand remains constant for each class; consequently, the number of classes must be increased significantly in order to draw a curve that is as close to reality as possible. This study overcomes this limitation by proposing a polynomial interpolation [8] in each class, as obtained by the Lagrange method that provides an efficient algorithm to find the interpolating polynomial. This method also helps finding the interpolated value for a specific x without the need to identify the analytical expression of the interpolating polynomial by using the expression $p(x) = y_0 L_0(x) + y_1 L_1(x) + \cdots + y_n L_n(x)$, where:

$$L_{n,k}(x) = \prod_{\substack{i=0 \\ i \neq k}}^{n} \frac{x - x_i}{x_k - x_i} \tag{11.4}$$

In conclusion, the frequency tables that result from the modeling of the load curve can contain solely 1st and 2nd degree interpolating polynomials. For the linear case, only the end nodes are available; and for the quadratic case, not only the end points, but also the mid-point, are available. In this fashion, the functional relation between a specific level of demand and its associated probability of occurrence may be obtained.

11.2.1.3 Probabilistic Indicators

Loss of Load Probability

The Loss of Load Probability (LOLP) reliability indicator defined in Eq. (11.5) indicates the probability that a particular level of demand may not be satisfied with the available generation capacity. This indicator is defined as the number of days or hours per year when the available generating capacity is not enough to satisfy the energy requirements for all the daily or hourly loads. LOLP is usually expressed as a time ratio; for example, 0.1 days per year, which equals to a probability of 0.000274 (that is, 0.1 / 365) [4].

LOLP is currently the most widely used indicator, and it is also perhaps the most misunderstood, due to the incorrect use of the term probability in its name. In an attempt to clarify this confusion, Billinton [2] has defined the term LOLE as the number of days (or hours) per year in which the available generation capacity is not enough to supply the required energy to the maximum daily (or hourly) load. Consequently, LOLP is redefined as LOLE / N, where N is the number of time increments in the calculated LOLE (N $=$ 365, if LOLE is calculated from maximum daily load data and expressed in terms of days, while N $=$ 8760, if LOLE is calculated from hourly load data).

$$LOLP = \sum_{i \in S} P_i \qquad (11.5)$$

$$LOLE = LOLP \times 8760 \qquad (11.6)$$

where P_i is the probability that the system is in the i state and S represents the set of all the load loss states of the system.

Expected Energy Not Supplied

EENS, as defined in (11.7), measures the expected amount of energy that will not be supplied per year due to deficiencies in the generation capacity and/or shortage of basic energy supply.

Mathematically, EENS (expressed in units MWh/year) is the sum of the weighted probability of power outages caused by deficient capacity throughout a year. This indicator is widely used in Europe, where it is one of the most common indicators relied upon to evaluate generation performance reliability. EENS is very useful for public utilities that operate such limited energy technologies as hydro, solar, and wind.

$$EENS = \sum_{i \in S} C_i \cdot F_i \cdot D_i = \sum_{i \in S} 8760 \times C_i \cdot P_i \qquad (11.7)$$

$$F_i = P_i \sum_{k \in N} \lambda_k \qquad (11.8)$$

where C_i is the disconnected demand of the system in the i state, D_i is the time that the system remains in the i state, and F_i is the state frequency of the system defined in Eq. (11.8). F_i can be calculated as the relation between frequency and state probability of the system P_i, where λ_k is the output ratio of the k component, and N is the set of all the system components.

11.2.1.4   Considerations for the MCNS Simulation Method

An MCNS simulation method, as applied to HL-I, is completed after the above-mentioned 4 Steps have been run and the following steps are executed:

Step 5: A random uniform number U_i is generated to obtain a demand level i, as was previously defined when modeling a load curve.

Step 6: The above-defined system reliability indicators are stored, and Steps 1–6 are repeated until the stopping criterion [5] is reached.

This procedure helps obtain the proposed indicators for the system discussed herein, as well as for any system subjected to the effects of similar conditions.

11.2.2 Models and Methods for the Security Assessment

11.2.2.1 Power Flow

Each EPS component has a distinct representation in load flow studies. The generation is represented by the active power produced by a generator and the reactive power consumed and produced by a generator. Transformers and transmission lines are represented by their Π nominal circuit. Loads are always represented by the demand leaving the node, if inductive. The static and synchronous capacitors are represented by their generation or consumption of reactive power.

The most convenient method for solving power flow (PF) in an EPS is the node current method. It is simpler to simulate the member elements of the power grid. This study applies the Newton–Raphson (NR) method to solve load flow equations. Most load flow programs use the power equations in polar form [16]. For this reason, this study applies them too.

The equation for the full power injected into a node k of an EPS of n nodes is

$$\underline{S}_k = \underline{V}_k \cdot \underline{I}_k^* = P_k + jQ_k \tag{11.9}$$

The equation for the current injected into a node k can be expressed as:

$$\underline{I}_k = \underline{Y}_{k1} \cdot \underline{V}_1 + \underline{Y}_{k2} \cdot \underline{V}_2 + \cdots + \underline{Y}_{kn} \cdot \underline{V}_n = \sum_{i=1}^{N} \underline{Y}_{ki} \cdot \underline{V}_i \tag{11.10}$$

By substituting (11.10) in (11.9), and with the usual nomenclature for the elements of the admittance matrix,

$$P_k + jQ_k = \underline{V}_k \sum_{i=1}^{N} \underline{V}_i^* \cdot (G_{ki} - jB_{ki}) \tag{11.11}$$

where $\underline{V}_i^*$ and $G_{ki} - jB_{ki}$ are the conjugated $\underline{V}_i$ and $\underline{Y}_{ki}$, respectively.

The product of the two voltages in Eq. (11.11) can be developed as a Taylor series expansion of real numbers, and the real and imaginary parts of Eq. (11.11) must be separated and expressed as:

$$P_k = V_k \sum_{i=1}^{N} V_i \cdot (G_{ki} \cos(\delta_{ki}) + B_{ki} \sin(\delta_{ki})) \tag{11.12}$$

$$Q_k = V_k \sum_{i=1}^{N} V_i \cdot (G_{ki} \sin(\delta_{ki}) - B_{ki} \cos(\delta_{ki})) \qquad (11.13)$$

where V_k and V_i modules of the voltages in nodes k and i, respectively, per unit, and δ_k and δ_i angles of the voltages in nodes k and i, respectively, in degrees, and $\delta_{ki} = \delta_k - \delta_i$ subtraction angles of the voltages at nodes k and i.

Equations (11.12) and (11.13) show that the net active and reactive powers of each node are functions of the modules and the angles of the voltages on all nodes of the EPS. In the load nodes, both generation and demand are specifically determined; therefore, the net powers are established. By substituting these new variables, the load flow equations, as expressed in symbolic notation by the Newton–Raphson method, are

$$\begin{bmatrix} (\Delta P) \\ (\Delta Q) \end{bmatrix} = \begin{bmatrix} \frac{\partial(P)}{\partial(\delta)} & \frac{\partial(P)}{\partial(|V|)} \\ \frac{\partial(Q)}{\partial(\delta)} & \frac{\partial(Q)}{\partial(|V|)} \end{bmatrix} \cdot \begin{bmatrix} \Delta\delta \\ \Delta|V| \end{bmatrix} = \begin{bmatrix} J_1 & J_2 \\ J_3 & J_4 \end{bmatrix} \cdot \begin{bmatrix} \Delta\delta \\ \Delta|V| \end{bmatrix} \qquad (11.14)$$

where: J_1, J_2, J_3 y J_4 are Jacobian matrices of the above equations system, which show the changes in the active and reactive powers when the module and the angle of the voltage change. Likewise, (ΔP) and (ΔQ) are column vectors representing the differences between the full power that actually exists at each node and the power that is calculated from the voltage of each iteration. These differences are called power mismatch. The convergence of the Newton–Raphson method is obtained from this mismatch.

Under this method, small values of tolerance are generated, because the resulting reference is the documented actual net power. If (11.14) is expressed the same way as (11.15), the tension in the nodes may be calculated using (11.16) as follows, since in the PQ nodes, the unknowns are the tensions:

$$[\Delta S] = [J] \cdot [\Delta V] \qquad (11.15)$$

$$[\Delta V] = [J]^{-1} \cdot [\Delta S] \qquad (11.16)$$

11.2.2.2 Continued Power Flow

The continuous load flows method is widely applied to the reliability assessment of steady-state regimes, in which the system load is stepped up to a point where the corresponding load flow ceases to converge, thus indicating that the system has reached an unstable state. The application of the aforementioned method takes into account the system's operating restrictions, such as reactive power generation limits in PV nodes, voltage limits in loading nodes, and transfer limits on lines [16].

The Jacobian matrix of the Eq. (11.14) becomes singular in the determination of voltage stability limit. Consequently, conventional power-flow algorithms are prone to convergence problems in operating conditions close to the stability limit.

Fig. 11.1 Predictor step
obtained by the tangent vector

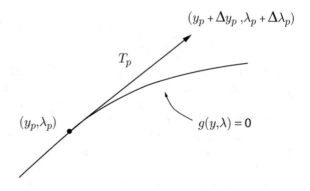

The continuous power-flow determination can overcome this problem by reformulating the power-flow equations, in a manner that all the possible load regimes remain properly conditioned for continued V-P curve charting. The above solves both the power-flow problem for stable state and the unstable equilibrium point problems (i.e., both the top and bottom of the V-P curve). The continuous power-flow method discussed in this section was obtained from [14].

The continued load flow method consists of a first step, tangent to the function, executed at the point that corresponds to the initial flow (predictor step), regardless the active and reactive power balance restrictions, as well as corrective step, perpendicular to the initial flow, which drives the system to conform to the aforementioned restrictions. Figure 11.1 graphically illustrates the above.

In order to implement the predictor step in any generic point of balance, the following relation may be applied:

$$g(y_p, \lambda_p) = 0 \Rightarrow \left.\frac{dg}{d\lambda}\right|_p = 0 = \left.\nabla_y g\right|_p \frac{dy}{d\lambda} + \frac{dg}{d\lambda} \tag{11.17}$$

and the tangent vector can be approximated by:

$$\tau_p = \left.\frac{dg}{d\lambda}\right|_p \approx \frac{\Delta y_p}{\Delta \lambda_p} \tag{11.18}$$

From Eqs. (11.17) and (11.18), the following is obtained:

$$\tau_p = -\nabla_y g\Big|_p^{-1} \left.\frac{dg}{d\lambda}\right|_p \tag{11.19}$$

$$\Delta y_p = \tau_p \Delta \lambda_p$$

If a constant k is selected as control for the step size, then Δy_p and $\Delta \lambda_p$ increases can be determined, and standardization may help to avoid an extensive step, when $|\tau_p|$ is large.

$$\Delta \lambda_p \triangleq \frac{k}{|\tau_p|} \quad \Delta y_p \triangleq \frac{k\tau_p}{|\tau_p|} \tag{11.20}$$

where $k = \pm 1$, and determines the increase or decrease of λ.

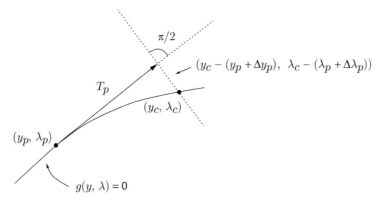

Fig. 11.2 Step correction obtained by the perpendicular intersection

In the corrector step, a set of $n + 1$ equations are solved, as follows:

$$g(y, \lambda) = 0 \tag{11.21}$$

$$\rho(y, \lambda) = 0$$

where solution g must be multiple bifurcation and ρ is the additional equation that guarantees the non-singularity of the system at its turning point. In the case of ρ, two options are available: perpendicular intersection and local parametrization.

In the case of perpendicular interception (see Fig. 11.2), ρ is expressed as follows:

$$\rho(y, \lambda) = \begin{bmatrix} \Delta y_p \\ \Delta \lambda_p \end{bmatrix}^T \begin{bmatrix} y_c - (y_p + \Delta y_p) \\ \lambda_c - (\lambda_p + \Delta \lambda_p) \end{bmatrix} = 0 \tag{11.22}$$

For the local parametrization, the λ parameter and/or the y_i variable are forced to a fixed value:

$$\rho(y, \lambda) = \lambda_c - \lambda_p - \Delta \lambda_p \tag{11.23}$$

$$\rho(y, \lambda) = y_{ci} - y_{pi} - \Delta y_{pi}$$

The selection of the variable depends on multiple bifurcation g, as described in Fig. 11.3.

11.2.2.3 Modal Analysis

The power industry has largely depended on conventional energy-flow software packages for its static analysis of voltage stability, as determined by calculating the P-V and Q-V curves in selected load nodes. Generally, such curves are generated

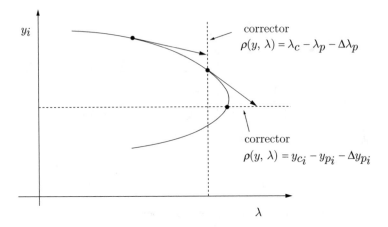

Fig. 11.3 Step correction obtained through local parametrization

by running a large number of power flows. Such procedures can be automated; however, they are time consuming, and fail to provide all the necessary information to identify the causes of any stability problems. Moreover, these procedures focus on individual nodes; that is, stability characteristics are set for each node independently. As a result, the stability condition of the system may appear distorted. The choice of nodes for Q-V and P-V curve assessments must be made carefully, and a large number of these curves are required for complete information. In fact, it may not be possible to generate Q-V curves completely due to the divergence of power flows caused by problems elsewhere in the system; therefore, a modal analysis is required as a supplement to voltage stability studies.

The modal analysis, as described in [14], has been applied to voltage stability assessment of actually operating systems. The advantages of this approach include its supply of overall voltage stability information about the system, and its contribution to clearly identify potentially troubling areas.

As shown previously, the network restrictions can be expressed in a linearized form as follows:

$$\begin{bmatrix} \Delta P \\ \Delta Q \end{bmatrix} = \begin{bmatrix} J_{P\delta} & J_{PV} \\ J_{Q\delta} & J_{QV} \end{bmatrix} \cdot \begin{bmatrix} \Delta \delta \\ \Delta V \end{bmatrix} \tag{11.24}$$

where ΔP represents gradual changes in the active power at the node, ΔQ represents gradual changes in the reactive power injected into the node, $\Delta \delta$ represents gradual changes in the angle of the voltage at the node, and ΔV represents gradual changes in the magnitude of the voltage at the node.

The terms of the EPS Jacobian matrix in Eq. (11.24) are modified by $J_{P\delta}$, J_{PV}, $J_{Q\delta}$, and J_{QV} to form the system's Jacobian matrix. The stability of the system voltage is affected by both P and Q. However, at each operating point, P can be kept constant, and the voltage stability may be assessed by calculating the incremental ratio between Q and V. This approach is analogous to the V-Q curve. Although incremental changes in P are neglected for the purpose of the formulation, the effects of changes in the system load and/or energy transfer are taken into consideration by evaluating the gradual relation between Q and V in different operating conditions. Based on the foregoing considerations, in Eq. (11.24), when $\Delta P = 0$, then:

$$\Delta Q = \left[J_{QV} - J_{Q\delta} \cdot J_{P\delta}^{-1} \cdot J_{PV} \right] \tag{11.25}$$

where J_R is the system's reduced Jacobian Matrix.
From Eq. (11.25), we can write:

$$\Delta V = J_R^{-1} = \Delta Q \tag{11.26}$$

The stability characteristics of the system voltage can be identified by calculating the eigenvalues and eigenvectors of the reduced Jacobian matrix J_R defined by the Eq. (11.25). When:

$$J_R = \xi \cdot \Lambda \cdot \eta \tag{11.27}$$

where ξ represents the right eigenvector matrix, η the left eigenvector matrix, and Λ diagonal eigenvalues matrix of J_R.
For Eq. (11.27), $J_R^{-1} = \xi \cdot \Lambda^{-1} \cdot \eta$, and substituting in Eq. (11.26) the following is obtained:

$$\Delta V = \xi \cdot \Lambda^{-1} \cdot \eta \cdot \Delta Q \tag{11.28}$$

$$\Delta V = \sum_i \frac{\xi_i \cdot \eta_i}{\lambda_i} \cdot \Delta Q$$

where ξ_i is the ith column-right eigenvector and η_i the ith row-left eigenvector of J_R. Each eigenvalue λ_i and its corresponding eigenvector right and left ξ_i and η_i defines the ith mode Q-V response. Since $\xi^{-1} = \eta$, Eq. (11.28) can be written as:

$$\eta \cdot \Delta V = \Lambda^{-1} \cdot \eta \cdot \Delta Q \tag{11.29}$$

$$v = \Lambda^{-1} \cdot q$$

where $v = \eta \cdot \Delta V$ is the vector of modal voltage variation, and $q = \eta \cdot \Delta Q$ is the vector of modal reactive power variation.
The difference between Eqs. (11.26) and (11.29) is that Λ^{-1} is a diagonal matrix where J_R^{-1}, in general, is not diagonal. Equation (11.29) represents uncoupled first-order equations. Therefore, for the ith mode we have

$$v_i = \frac{1}{\lambda_i} \cdot q_i \tag{11.30}$$

If $\lambda_i > 0$, the ith modal voltage and the ith modal reactive power variations are along the same direction, indicating that the system voltage is stable. If $\lambda_i < 0$, it indicates that the system voltage is unstable. The magnitude of λ_i determines the degree of stability of the ith modal voltage. The smaller the magnitude of positive λ_i, the closer the ith modal voltage is to being unstable. When $\lambda_i = 0$, the ith modal voltage collapses, as any change in this modal reactive power causes an infinite change in modal voltage. The magnitude of the eigenvalues can provide a relative measure of proximity to instability; however, the eigenvalues do not provide an absolute measure due to the nonlinearity of the problem. The application of modal analysis helps determine how stable a system is, as well as the load and/or power that may added and/or transferred. When the system reaches its critical voltage stability point, the modal analysis helps identify the critical areas of voltage stability and the participating elements in each mode.

11.2.3   Methodology

For its proper planning and operation, an EPS must be reliable. Conceptually, reliability is composed of two attributes: adequacy and safety. Both attributes must be considered in order to arrive to more comprehensive conclusions. Planning requires standards or benchmarks for the comparison and evaluation of the performance of the operating EPS and the validation of the applied methodology. This chapter proposes the methodology described in Fig. 11.4 to assess the reliability of an EPS.

11.2.3.1   Adequacy Analysis

The factors that may be included in the assessment of the adequacy of an EPS system are many. This chapter assesses a generation system; therefore, inter alia, the following factors are reviewed: generating units in reduced power states, failing generator components, and random demand behavior. For the purpose of a clear understanding of possible results, it is important that the conditions discussed below are met.

Fig. 11.4 Proposed conceptual framework

1. The adequacy indicators are calculated on an equivalent bar, where all the generators are assumed to be in parallel, and the total system demand is the sum of all the demands from the total load nodes; in other words, only the generators (HL-I) are being considered.
2. The mathematical modeling of the generators is based on a two-state model.

The EPS indicators support an overall assessment of its adequacy. Nevertheless, for proper results, it is critical that the number of samples for the MC simulation be carefully selected. The criterion supporting this chapter is discussed in [6] for the MCNS simulation.

This study applies mainly the LOLE and EENS parameters to the systems discussed herein. These indicators are the most widely used in the world as benchmarks. Based on the estimates values produced by these indicators and other aforementioned considerations, an EPS may be assessed as adequate or inadequate.

11.2.3.2   Security Analysis

The safety analysis discussed herein focuses on the generation–transmission system (HL-II). The performance of the EPS voltage stability is assessed against certain contingencies, including but not limited to loss of system elements, such as lines and generating units. As this case study denotes, an EPS that supports the contingencies addressed herein is deemed safe in respect of its voltage stability.

To reach the above conclusions, this study proposes the following steps:

1. Identify the node(s) where the EPS under review is weak from the perspective of voltage stability. For this purpose, modal analysis helps to determine, from an overall standpoint, the nodes and/or bars that are closer to the point of voltage stability loss.
2. Run a continued power-flow diagram that helps chart P-V curves as supplementary information to the modal analysis.
3. Apply the contingencies proposed for the EPS, and illustrate the performance of eigenvalues, load parameter λ, and P-V curves for each case under review.

11.3   Result Analysis and Discussion

This chapter uses the Roy Billiton Test System (RBTS) [3]. Its purpose is to provide a common testing system for the design of different assessment procedures, the results of which may be reliably compared.

RBTS is a small system developed by the University of Saskatchewan for academic purposes. This system consists of six nodes, including five load nodes. It also has eleven generators, nine transmission lines, and seven links. Its total installed capacity is 240 MW, with a 185 MW peak demand. The voltage level for the entire system is 230 kV, and its mono-line is shown in Fig. 11.5.

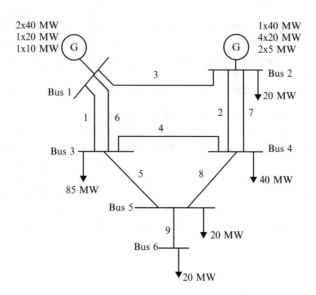

Fig. 11.5 RBTS mono-line system diagram

RBTS uses the demand model proposed for IEEE RTS, as discussed in [13]. Under this model, the hourly demand can be modeled; i.e., 8760 h a year are modeled as percentages of the maximum demand of the EPS under review.

11.3.1 Analysis of the Adequacy of the RBTS

Previous studies [6] show that adequacy ratios can be estimated with acceptable accuracy using two million samples in an RBTS. Notwithstanding, this paper was based on the well-established view discussed in [6], according to which, the number of samples required is much higher; in fact, fivefold or ten million samples.

The methods applied in this study require, to a certain extent, their validation. For this reason, these methods have been compared with procedures found in available literature. For the purpose of this comparison, reference [10] was chosen for its results on a HL-I test using RBTS.

Reference [10] uses the analytical method described in [4], the MERCORE program developed by the University of Saskatchewan, and the Monte Carlo Sequential simulation approach (MCS) addressed in [6].

This chapter applies the MCNS simulation method to calculate probabilistic adequacy ratios that result from modeled generator states and the chronological behavior of load curves. Table 11.1 shows the results for peak demand (assuming that the system operates at maximum demand all year round). It was found that all these methods supply similar estimates for the EPS under review. In the case of the

Table 11.1 Peak-demand analysis

Index	Analytical method [10]	MERCORE [10]	SMCS [6]	SMCNS
LOLE (h/year)	73.07	72.49	70.51	73.3965
EENS (MWh/year)	823.26	817.23	812.82	838.89

Table 11.2 Chronological demand performance analysis

Index	Analytical method [10]	MERCORE [10]	SMCS [6]	SMCNS
LOLE (h/year)	1.09	1.15	1.07	1.1546
EENS (MWh/year)	9.86	11.78	9.65	9.8680

proposed MCNS simulation method, the relative errors from the LOLE and EENS indicators are 0.0034 and 0.0047, respectively.

The EPS indicators are typically calculated using an annualized (8760 h) chronological load model. This paper uses the load duration curve defined in [13], as also applied by Gao [10].

The load curve model exercises considerable influence over the estimated indicators, as evinced by miscalculations. In this paper, a 52-class cumulative frequency histogram is used as a model. In addition to class intervals, a linear interpolation is performed to reflect the actual load curve as much as possible. Table 11.2 shows the rates of a HL-I system, as obtained from the aforementioned models, where the relative error for the LOLE and EENS indicators in the proposed MCNS simulation method are 0.0275 and 0.0387, respectively.

The results obtained using the MCNS simulation method are similar to those obtained with the analytical method and the exact model (8760 h). The main conclusion to be drawn from Table 11.2 is that the load model is the main reason for the variability in rates of reliability of the system.

11.3.2 Analysis of the Security of RBTS

As shown in Table 11.3 the system is stable because all its eigenvalues are positive. Nevertheless, as its eigenvalues drop, the voltage comes closer to becoming unstable. The eigenvalue represents a relative measure of proximity to instability. The results in Table 11.3 show that Node 6 is the most influential node over the two eigenvalues are closer to zero; therefore, Node 6 is the most sensitive to voltage stability. Figure 11.6 confirms the aforementioned claim. As the demand for power load rises, Node 6 visibly determines the voltage stability limit, since the voltage drop is greater as the active power of the EPS rises.

After Node 6 has been established as the most sensitive, the contingencies reflected in Table 11.4 helps to describe the performance of the eigenvalues, as well as the λ top-load and operational parameters, as shown in Fig. 11.7. It can be concluded that the most severe contingency for the RBTS occurs when a 40 MW

Table 11.3 Base-case
eigenvalues obtained from
RBTS modal analysis

Eigenvalues	Real part	Imaginary part	Node
1	36.9955	0	3
2	32.4651	0	4
3	2.3903	0	6
4	11.4311	0	6
5	999	0	1
6	1998	0	2

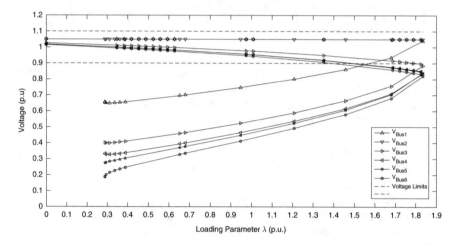

Fig. 11.6 Base-case P-V curve for RBTS

Table 11.4 RBTS contingency analysis

Contingencies analyzed	Eigenvalues (real part)	Maximum parameter λ (p.u.)	Operation parameter λ (p.u.)
Without contingencies	1	2	3
Generator of 40 MW	4	5	6
Generator of 40 MW, line 1–2	7	8	9
Generator of 40 MW, line 1–3	10	11	12
Generator of 40 MW, line 2–4	13	14	15
Generator of 40 MW, line 3–5	16	17	18

generator and the 1–2 transmission line are down. Figure 11.8 shows the P-V curves
for the case in discussion. While subjected to severe contingencies, the system under
review is apparently able to absorb incremental in demand; i.e., the system may even
take into account the uncertainty that may exist in its forecast demand.

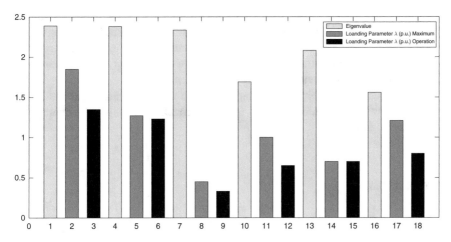

Fig. 11.7 Performance of the contingencies assessed for Node 6 of the RBTS

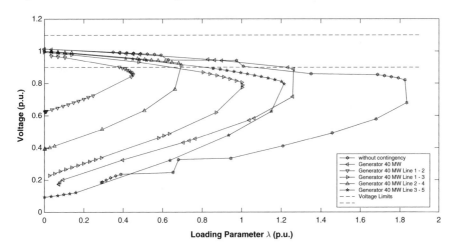

Fig. 11.8 P-V curve for the contingencies assessed at Node 6 of the RBTS

11.4   Conclusions and Recommendations

The methodology applied to Node 6 of the RBTS helps to assess the reliability of the EPS under review in specific conditions, with regard to its conceptual components; i.e., adequacy and safety.

The adequacy evaluation using the LOLE and EENS probabilistic indicators shows that the RBTS is suitable: Its LOLE equals 1.1546 h per annum; that is, below the desirable 2.4 h per annum rate used as a probabilistic criterion for acceptable risk in an EPS.

The results from previous assessments of various indicators in static methods, that mathematically model the voltage performance in an EPS, have consistently shown that the RBTS discussed here is safe in the context of its tested contingencies. It has also been demonstrated that Node 6 is the weakest point; a claim consistent with the results from previous works.

The lessons learned from this chapter lead to the conclusions that the RBTS under review is reliable. The results from this chapter may represent appropriate EPS planning tools.

References

1. Bansal, R.C., Bhatti, T.S., Kothari, D.P.: Discussion of bibliography on the application of probability methods in power system reliability evaluation. IEEE Trans. Power Syst. **17**(3), 924–935 (2002)
2. Billinton, R.: Considerations in including uncertainty in LOLE calculations for practical systems. In: IEEE Paper A 79 075-3 (presented at the Power Engineering Society Winter Meeting, New York) (1979)
3. Billinton, R.: A reliability test system for educational purposes basic data. IEEE Trans. Power Syst. **PWR-3**(4), 1238–1244 (1989)
4. Billinton, R., Allan, R.N.: Reliability Evaluation of Power Systems, 2nd edn. Plenum, New York (1996)
5. Billinton, R., Li, W.: Reliability Assessment of Electrical Power Systems Using Monte Carlo Methods. Plenum, New York (1994)
6. Billinton, R., Wangdee, W.: Impact of utilizing sequential and non-sequential simulation techniques in bulk electric system reliability assessment. IEEE Gener. Transm. Distrib. **152**(5), 623–628 (2005)
7. Billinton, R., Fotuhi-Firuzabad, M., Bertling, L.: Bibliography on the application of probability methods in reliability evaluation 1996–1999. IEEE Trans. Power Syst. **16**(4), 595–602 (2001)
8. Cheney, W., Kincaid, D.: Numerical Mathematics and Computing, 6th edn. Thomson Corporation, USA, (2004)
9. Fuentes, V., Duarte:, O.: Evaluación mediante enumeración de estados de la confiabilidad del Sistema Interconectado del Norte Grande de Chile (SING). Revista Chilena de Ingeniería **19**(2), 292–306 (2011)
10. Gao, Y.: Adequacy assessment of electric power systems incorporating wind and solar energy. Master's thesis, University of Saskatchewan (2006)
11. García, S., González, J., Boza, J.: Aplicación de flujos de cargas sucesivos con Jacobiana constante para la determinación del punto de colapso de tensión. Validación con patrón IEEE-14 nodos. Ingeniería Energética **XXXII**, 43–55 (2011)
12. Hernández, N.B.V.: Load forecast uncertainty considerations in bulk electrical system adequacy assessment. Master's thesis, University of Saskatchewan (2009)
13. IEEE Task Force: IEEE reliability test system. IEEE Trans. Power Apparatus Syst. **PAS-98**(6), 2047–2054 (1979)
14. Kundur, P.: Power System Stability and Control. McGraw-Hill, New York (1994)
15. Lange, K.: Numerical analysis for statisticians. Statistics and Computing. Springer, New York (2010)
16. Saadat, H.: Power System Analysis. McGraw-Hill, New York (1999)

Chapter 12
Polymeric Thin Film Transistors Modeling in the Presence of Non-Ohmic Contacts

Magali Estrada del Cueto, Antonio Cerdeira Altuzarra, Benjamín Iñiguez Nicolau, Lluis F. Marsal Garvi, and Josep Pallarés Marzal

Abstract This chapter discusses a model for polymeric thin film transistors, PTFTs, based on the unified model and parameter extraction method (UMEM), previously developed by these authors for thin film transistors, which includes specific features of OTFTs, as initial drain current and subthreshold behavior. In this case, the model shows the presence of non-ohmic contacts at drain and source. UMEM, has been previously used with a-Si:H, polysilicon and nanocrystalline TFTs and as compared with previous methods, it provides a more rigorous and accurate determination of the main electrical parameters of TFTs. Device parameters are extracted in a simple and direct way from the experimental measurements, with no need for assigning predetermined values to any model parameter or using optimization methods. In this case, the extraction procedure is complemented to include the extraction of the model parameters related to the non-ohmic contacts at drain and source. The model was implemented in Verilog-A and included for application in Spice simulators. Good consistency was found between the simulations in Spice using the model and the measured devices, as shown herein.

Keywords Thin film transistors • Series resistance • Polymeric transistors • Non-ohmic contacts • Compact modeling • OTFTs • Series resistance extraction

M.E. del Cueto (✉) • A.C. Altuzarra
Electrical Engineering Department, CINVESTAV-IPN, Av. IPN No. 2508, México, 07360, México
e-mail: mestrada@cinvestav.mx; cerdeira@cinvestav.mx

B.I. Nicolau • L.F.M. Garvi • J.P. Marzal
Department of Electronic, Electric and Automation Engineering, Universitat Rovira i Virgili, Avda. Paisos Catalans 26, Tarragona, 43007, Spain
e-mail: benjamin.iniguez@urv.cat; lluis.marsal@urv.cat; josep.pallares@urv.cat

© Springer International Publishing Switzerland 2016
A.J. da Silva Neto et al. (eds.), *Mathematical Modeling and Computational Intelligence in Engineering Applications*, DOI 10.1007/978-3-319-38869-4_12

12.1 Introduction

The simulation of electronic devices requires compact models that accurately reflect the behavior of the devices. At the same time, the model must include a specific extraction procedure to determine, in a simple and precise manner, model parameters. These models are introduced in commercial simulators for use in circuit design.

Many compact models have been developed for thin film transistors (TFTs) including but not limited to the RPI model based on Shur's work [8], which is widely used. For organic TFTs (OTFTs), several models have also been developed [4–7, 11]. Using the representation of the mobility as a power dependence of the gate voltage, which is a typical behavior of the amorphous TFTs, the unified model and extraction method (UMEM) was developed, for which a simple and specific extraction procedure for model parameters was included [1].

This model and its extraction procedure were first applied to a-Si:H TFTs, to nanocrystalline TFTs, and some types of polycrystalline TFTs. More recently, the model was complemented for use with organic TFTs (OTFTs) including polymeric TFTs, PTFTs. For this last application, it was necessary to consider frequent effects present in these devices, such as the initial current between drain (D) and source (S) and the behavior in subthreshold [2]. Since there are no PN junctions at D and S, the material conductivity is usually high enough to give rise to a current between D and S even when the gate voltage (V_{GS}) is below the threshold voltage (V_T).

Another specific issue associated with TFTs which must be taken into account in their modeling is the effect of the series resistance. In these devices, the current from the D contact has to flow across the semiconductor layer width, in order to reach the device channel. Similarly, the channel current has to flow across the semiconductor layer width, in order to reach the S contact.

Finally, since the value of the energy corresponding to the highest occupied molecular orbit, which is the equivalent in organic materials to the valence band in the inorganic materials, is greater than 5 eV, most metals form non-ohmic contacts at D and S. These non-ohmic contacts deform the origin of the output characteristics of the OTFTs and this effect has to be included in the model. In [10], a first approach to model this effect required a considerable number of empiric parameters, which had to be calculated by optimization procedures. UMEM reduced the number of parameters required and used a more physically based representation; but still, the extraction of the parameters was complicated [2]. This chapter presents a new proposal for modeling and extracting the model parameters to represent the effect of the non-ohmic contacts. As compared with its predecessors, this new proposal offers several advantages.

12.2 Basic Aspects of UMEM

12.2.1 Basic Expressions to Represent the Above Threshold Operating Region

As already mentioned, the charge mobility in TFTs is usually represented as depending on the gate voltage elevated to a power [3]:

$$\mu_{\text{FET}} = \frac{\mu_{oo}}{V_{\text{AA}}^{\gamma}}(V_{\text{GS}} - V_T)^{\gamma} \tag{12.1}$$

where μ_{oo} is taken as 1 cm²/Vs for dimensional purposes, γ describes the variation of mobility as a power function of V_{GS}. The low field mobility, when $(V_{\text{GS}} - V_T) = 1$ is written as:

$$\mu_{\text{FETo}} = \frac{\mu_{oo}}{V_{\text{AA}}^{\gamma}} \tag{12.2}$$

The drain current as function of the drain voltage V_{DS} and V_{GS} is written as [3]:

$$I_{\text{DSa}} = \frac{W}{L}C_i\mu_{\text{FET}}\frac{(V_{\text{GS}} - V_T)V_{\text{DS}}(1 - \lambda V_{\text{DS}})}{\left[1 + R\frac{W}{L}C_i\mu_{\text{FET}}(V_{\text{GS}} - V_T)\right]\left\{1 + \left[\frac{V_{\text{DS}}}{\alpha_S(V_{\text{GS}} - V_T)}\right]^m\right\}^{\frac{1}{m}}} \tag{12.3}$$

where W and L are the width and channel length of the transistor, respectively, R is the series resistance, C_i is the capacitance per unit area of the gate dielectric, m allows modeling the transition of the output characteristic from the linear to the saturation region, λ describes the slope or the output curve in the saturation region, and α_s is the saturation coefficient, which is determined as:

$$V_{\text{DSsat}} = \alpha_S(V_{\text{GS}} - V_T) \tag{12.4}$$

When the drain current behaves as:

$$I_{\text{DS}} \propto (V_{\text{GS}} - V_T)^n \tag{12.5}$$

The properties of the integral function $H1$ can be used:

$$H1(V_{\text{GS}}) = \frac{\int_0^{V_{\text{GSmax}}} I_{\text{DS}}(V_{\text{GS}})\,dV_{\text{GS}}}{I_{\text{DS}}(V_{\text{GS}})} \tag{12.6}$$

Calculating $H1(V_{\text{GS}})$ for the drain current in the linear region $I_{\text{DSlin}}(V_{\text{GS}})$, when the drain voltage is equal to V_{DS1}:

$$H1(V_{\text{GS}}) = \frac{\int_0^{V_{\text{GSmax}}} I_{\text{DSlin}}(V_{\text{GS}})\,dV_{\text{GS}}}{I_{\text{DSlin}}(V_{\text{GS}})} = \frac{1}{2 + \gamma}(V_{\text{GS}} - V_T) \tag{12.7}$$

Using this expression, the following steps are used to extract model parameters:

Step 1: Calculate the slope and intercept of Eq. (12.7), where:

$$\gamma = \frac{1}{slope} - 2 \tag{12.8}$$

$$V_T = \frac{intercept}{slope} \tag{12.9}$$

Step 2: From the equation $y(V_{GS})$ described below:

$$y(V_{GS}) = I_{DS_{lin}}(V_{GS})^{\frac{1}{1+\gamma}} = PA(V_{GS} - V_T) \tag{12.10}$$

determine the slope PA, which will be equal to:

$$PA = \left[\frac{\left(\frac{W}{L}\right) C_i \, \mu_{oo} \, V_{DSI}}{V_{AA}^{\gamma}} \right]^{\frac{1}{1+\gamma}} \tag{12.11}$$

and V_{AA} is determined from Eq. (12.11).

Step 3: Calculate the low field mobility value μ_{FETo} as:

$$\mu_{FETo} = \frac{\mu_{oo}}{V_{AA}^{\gamma}} = \left[\frac{PA^{1+\gamma}}{\left(\frac{W}{L}\right) C_i \, V_{DSI}} \right] \tag{12.12}$$

Step 4: Calculate the series resistance R as:

$$R = \frac{V_{DSI}}{I_{DSlin}(V_{GSmax})} - \frac{1}{\left(\frac{W}{L}\right) C_i \, \mu_{FET}(V_{GSmax} - V_T)} \tag{12.13}$$

Step 5: Using the transfer curve in saturation $I_{DSsat}(V_{GS})$, calculate the equation $y(V_{GS})$ described below:

$$y(V_{GS}) = I_{DSsat}(V_{GS})^{\frac{1}{1+\gamma}} = Ps(V_{GS} - V_T) \tag{12.14}$$

and calculate its slope Ps.

Step 6: Determine the saturation coefficient α_s as:

$$\alpha_S = \left[\frac{V_{DS2}}{PA^{1+\gamma}} \right] Ps^{2+\gamma} \sqrt{2} \tag{12.15}$$

where V_{DS2} is the value of V_{DS} at which the transfer curve in saturation was measured. It is recommended that the saturation transfer curve at the maximum drain voltage within the voltage operation range of the device be measured.

Step 7: Considering $\lambda = 0$ in Eq. (12.3), calculate m as:

$$m = \frac{\log 2}{\log \left[\dfrac{\frac{W}{L} C_i \left(\frac{\mu_{oo}}{V_{AA}^{\gamma}} \right) \alpha_S (V_{DSsat1})^{2+\gamma}}{I_{DSsat}(V_{DSsat1}) \left[1 + R \frac{W}{L} C_i \left(\frac{\mu_{oo}}{V_{AA}^{\gamma}} \right) \left(\frac{V_{DSsat1}}{\alpha_S} \right)^{1+\gamma} \right]} \right]} \tag{12.16}$$

where $V_{DSsat1} = \alpha_S(V_{GSI} - V_T)$ and the value for V_{GSI} is taken equal or near to the maximum value of V_{GS} applied to the transistor, V_{GSmax}.

Step 8: Calculate parameter λ as:

$$\lambda = \frac{\left[\frac{I_{DS2}}{V_{DS2}^2} \right] \left[1 + R \frac{W}{L} C_i \left(\frac{\mu_{oo}}{V_{AA}^{\gamma}} \right) (V_{GSI} - V_T)^{1+\gamma} \right] \left[1 + \left(\frac{V_{DS2}}{\alpha_S(V_{GSI}-V_T)} \right)^m \right]^{\frac{1}{m}}}{\frac{W}{L} C_i \left(\frac{\mu_{oo}}{V_{AA}^{\gamma}} \right) (V_{GSI} - V_T)^{1+\gamma}} - \frac{1}{V_{DS2}} \tag{12.17}$$

where the values for V_{GSI} and V_{DS2} are taken equal to or near the maximum values of the gate and drain voltages at which the device was measured. I_{DS2} is the measured drain current for V_{GSI} and V_{DS2}.

12.2.2   Basic Expressions to Represent the Subthreshold Operating Region

In the subthreshold, the drain current is represented as an exponential dependence with the gate voltage:

$$I_{DSb} = I_{DSo} \exp\left(-2.3 \frac{(V_{GS} - V_T)}{S} \right) \tag{12.18}$$

where I_{DSo} is the drain current evaluated at $V_{GS} = (V_T + DV)$, S is the subthreshold slope, and DV is an adjustment parameter fixing a value of V_{GS} near to V_T, which is taken as the starting point of the subthreshold region.

The total drain current is calculated as:

$$I_{DS} = |I_{DSa}| \left[\frac{1 + \tanh[V_{GS} - (V_T + DT)\,Q]}{2} \right]$$

$$+ |I_{DSb} + I_o| \left[\frac{1 - \tanh[V_{GS} - (V_T + DT)\,Q]}{2} \right] \tag{12.19}$$

where I_{DSa} is the current in above threshold regime and I_{DSb} is the current in subthreshold. I_o is the off current to be modeled and factor Q is a fitting parameter to sew the current in the two regions.

12.2.3 Modeling the Output Characteristics in TFTs with Non-Ohmic Contacts at D and S

When a non-ohmic contact resistance is present at D and S of the transistor, the applied voltage will fall partly across the transistor and partly across the resistance. This can be expressed as:

$$V_{DSext} + V_{DS} + V_{diode} = \frac{I_{DS}}{G(V_{GS}, V_{DS})} + n \frac{kT}{q} \log \left(\frac{I_{DS}}{I_{do}} \right) \qquad (12.20)$$

As already mentioned, in [2] the extraction procedure presented for the model parameters associated with the non-ohmic resistance was complicated and difficult to implement in Verilog-A. For this reason, the following new procedure is hereby proposed.

First, the output curve of the TFT corresponding to the highest applied gate voltage to the device is plotted in semilogarithmic scale.

Afterwards, the slope P_{diode} and intercept $interD$ in the region near the origin, where the curve is deformed, is determined. The ideality coefficient n of the rectifying contact can be determined as:

$$n = \frac{\log(e)}{kTP_{diode}} \qquad (12.21)$$

and the saturation current of the diode will be

$$I_{do} = 10^{interD} \qquad (12.22)$$

The nonlinear series resistance R_c can be written as:

$$R_c = R_{co} \exp(-\eta V_{DS}) \qquad (12.23)$$

where

$$R_{co} = \left[\frac{n k T \lg (I_{DS}(V_{GSmax}))}{I_{DS}(V_{GSmax})} \right] \qquad (12.24)$$

$$\eta = \frac{q}{nkT} \qquad (12.25)$$

The value of the total series resistance R_t will be

$$R_t = R + R_c \qquad (12.26)$$

The total contact resistance now substitutes resistance R in Eq. (12.3).

12.2.4 Summarizing the Steps to Extract Model Parameters when Non-Ohmic Contacts Are Present

When a nonlinear resistance is present, the procedure to calculate model parameters is the same up to Step 6. Prior to calculating parameter m, parameters n, I_{do}, and R_t must be calculated according to Eqs. (12.21)–(12.26). Subsequently, parameter m is calculated using Eqs. (12.16) and (12.17). The rest of the extracting procedure remains the same.

When a non-ohmic resistance is present, the determination of V_T from the linear transfer characteristic becomes inaccurate, since the values of V_{DS} at which the characteristic is to be measured, will fall in the deformed region. In this case, function H is applied to the transfer curve in saturation. In general, the modeling of devices that contain non-ohmic contacts is complicated; however, the procedure described herein generates quite acceptable results that can be applied to circuit design.

12.3 Examples of OTFT Characteristics Modeling

Figure 12.1 shows (a) linear transfer characteristics and (b) output characteristics, as measured and simulated in SmartSpice [9] for a pentacene OTFT. The simulation in SmartSpice was conducted using the aforementioned UMEM in Verliog-A as an external model.

Figure 12.2 shows the UMEN output characteristics measured and simulated in SmartSpice for a Poly (3-hexylthiophene), (P3HT) PTFT.

Figure 12.3 compares the measured and simulated output characteristics for a PTFT using poly(9,9-dioctylfluorenyl-co-bithiophene (F8T2), where the non-ohmic contact is present.

12.4 Conclusions

A new procedure for the extraction of model parameters applicable to the presence of a non-ohmic contact at D and S in OTFTs is described herein and incorporated to the general extraction procedure for model parameters of the UMEM. UMEM is implemented in Verilog-A. Good agreement was observed between measured and simulated in SmartSpice characteristics using UMEM as an external model.

Acknowledgements This chapter was supported by CONACYT Project 127978, Project TEC2011-28357-C02-01(UWBRFID), and Project TEC2012-34397 from the Ministerio de Economía y Competitividad de España, as well as project 2014-SGR-1344 from the Generalitat de Catalunya.

Fig. 12.1 Comparison of
measured and simulated in
SmartSpice using UMEM (**a**)
linear transfer characteristic;
(**b**) output characteristics for
a OTFT consisting of CYTOP
as dielectric and pentacene as
semiconductor

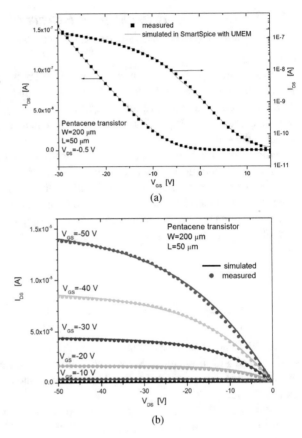

(a)

(b)

Fig. 12.2 Measured and
simulated in SmartSpice
using UMEM output
characteristics for a PTFT
using Polymethyl
methacrylate (PMMA) as
dielectric and
Poly(3-hexylthiophene),
(P3HT) as semiconductor

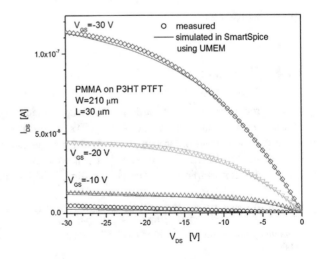

Fig. 12.3 Measured and simulated in SmartSpice using UMEM output characteristics for a PTFT with PMMA as dielectric and poly(9,9-dioctylfluorenyl-co-bithiophene) (F8T2) as semiconductor

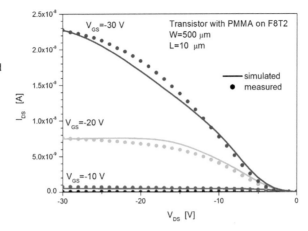

References

1. Cerdeira, A., Estrada, M., García, R., Ortiz-Conde, A., García, F.J.: New procedure for the extraction of basic a-si:h TFT model parameters in the linear and saturation regions. Solid State Electron. **45**, 1077–1080 (2001)
2. Estrada, M., Cerdeira, A., Puigdollers, J., Reséndiz, L., Pallares, J., Marsal, L., Voz, C., niguez, B.I.: Accurate modeling and parameter extraction method for organic TFTs. Solid State Electron. **49**(6), 1009–1016 (2005)
3. Fjeldly, T.A., Ytterdal, T., M.Shur: Introduction to Device Modeling and Circuit Simulation. Wiley, New York (1998)
4. Horowitz, G., Delannoy, P.: An analytical model for organics based thin film transistors. J. Appl. Phys. **70**, 469–475 (1991)
5. Marinov, O., Deen, M.J., Zschieschang, U., Klauk, H.: Organic thin film transistors. Part I Compact DC modeling. IEEE Trans. Electron Devices **56**, 2952–2961 (2009)
6. Mijalkovi, S., Green, D., Nejim, A., Rankov, A., Smith, E., Kugler, T., Newsome, C., Halls, J.: Uotft: Universal organic TFT model for circuit design. In: Digest of the 6th International Conference on Organic Electronics. Liverpool (2009)
7. Necliudov, P.V., Shur, M.: Modeling of organic thin film transistors of different designs. J. Appl. Phys. **88**, 6594–6597 (2000)
8. Shur, M., M'Hack: Physics of amorphous silicon based alloy field-effect transistors. J. Appl. Phys. **54**, 3831–3842 (1984)
9. SILVACO Inc: SmartSpice (2015)
10. Street, R., Salleo, A.: Contact effects in polymer transistors. Appl. Phys. Lett. **81**(15), 2887–2889 (2002)
11. Torsi, L., Dodabalapur, A., Katz, H.F.: An analytical model for short channel organic thin film transistors. J. Appl. Phys **78**, 1088–1093 (1995)

Printed in the United States
By Bookmasters